COUNCIL on FOREIGN

TASK FORCE REPORT NO. 83

U.S. ECONOMIC SECURITY

Winning the Race for Tomorrow's Technologies

Justin G. Muzinich, Gina M. Raimondo, and James D. Taiclet, *Chairs*
Jonathan E. Hillman, *Project Director*

The mission of the Council on Foreign Relations is to inform U.S. engagement with the world. Founded in 1921, CFR is a nonpartisan, independent national membership organization, think tank, educator, and publisher, including of *Foreign Affairs*. It generates policy-relevant ideas and analysis, convenes experts and policymakers, and promotes informed public discussion—all to have impact on the most consequential issues facing the United States and the world.

The Council on Foreign Relations takes no institutional positions on policy issues and has no affiliation with the U.S. government. All views expressed in its publications and on its website are the sole responsibility of the author or authors.

The Council on Foreign Relations sponsors Task Forces to assess issues of current and critical importance to U.S. foreign policy and provide policymakers with concrete judgments and recommendations. Diverse in backgrounds and perspectives, Task Force members aim to reach a meaningful consensus on policy through private deliberations. Once launched, Task Forces are solely responsible for the content of their reports. Task Force members are asked to join a consensus signifying that they endorse "the general policy thrust and judgments reached by the group, though not necessarily every finding and recommendation." Members' affiliations are listed for identification purposes only and do not imply institutional endorsement. Task Force observers participate in discussions but are not asked to join the consensus.

For further information about CFR or this Task Force, please write to the Council on Foreign Relations, 58 East 68th Street, New York, NY 10065, or call the Communications office at 212.434.9888. Visit our website, CFR.org.

Copyright © 2025 by the Council on Foreign Relations®, Inc.
All rights reserved.
Printed in the United States of America.

This report may not be reproduced in whole or in part, in any form beyond the reproduction permitted by Sections 107 and 108 of the U.S. Copyright Law Act (17 U.S.C. Sections 107 and 108) and excerpts by reviewers for the public press, without express written permission from the Council on Foreign Relations.

This report is printed on paper that is FSC® Chain-of-Custody Certified by a printer who is certified by BM TRADA North America Inc.

TASK FORCE MEMBERS

Task Force members are asked to join a consensus signifying that they endorse "the general policy thrust and judgments reached by the group, though not necessarily every finding and recommendation." They participate in the Task Force in their individual, not institutional, capacities.

Noubar Afeyan
Flagship Pioneering

Kevin Brown
Dell Technologies

Martin Chorzempa
Peterson Institute for
International Economics

Sarah Bauerle Danzman
Hamilton Lugar School,
Indiana University, Bloomington

Thomas Donilon
Former U.S. National Security Advisor

Elizabeth C. Economy
Hoover Institution, Stanford University

Edward B. Fishman
Center on Global Energy Policy,
Columbia University

Michèle A. Flournoy
WestExec Advisors

Jonathan E. Hillman
Council on Foreign Relations

Karen Karniol-Tambour
Bridgewater Associates

Aditi Kumar
Belfer Center for Science and
International Affairs,
Harvard Kennedy School

Sigal P. Mandelker
Ribbit Capital

Brent J. McIntosh
Citigroup, Inc.

Chris Miller
Fletcher School of Law and Diplomacy,
Tufts University

Craig Mundie
Mundie & Associates

Justin G. Muzinich
Muzinich & Co.

Nazak Nikakhtar
Wiley Rein LLP

Gina M. Raimondo
Council on Foreign Relations

Laura J. Richardson
U.S. Army, Retired

Peter L. Scher
JPMorganChase

Jonathan S. Spaner
McKinsey & Company

James D. Taiclet
Lockheed Martin Corporation

Laura D'Andrea Tyson*
Haas School of Business,
University of California, Berkeley

Sadek Wahba
I Squared Capital

Juan C. Zarate
Center on Economic and Financial
Power, Foundation for Defense
of Democracies

*The individual has endorsed the report and contributed an additional view.

CONTENTS

viii	*Foreword*
x	**EXECUTIVE SUMMARY**
02	**INTRODUCTION**
02	The Rise of Economic Security
04	The Race for Tomorrow's Technologies
09	China's Visible Hand
12	**WINNING THE RACE**
12	Artificial Intelligence
22	Quantum Technologies
28	Biotechnology
35	Critical Minerals
38	Workforce
40	**COMPETING IN THE AGE OF ECONOMIC WARFARE**
40	Establish an Economic Security Center
42	Upgrade Economic Security Tools
45	Guide Economic Security Actions With Principles
49	**APPENDIX**
49	Table I: Supply Chain Risks for AI
51	Table II: Supply Chain Risks for Quantum Technologies
54	Table III: Supply Chain Risks for Biotechology
55	*Additional Views*
56	*Endnotes*
64	*Acronyms*
66	*Acknowledgments*
67	*Task Force Members*
79	*Task Force Observers*
84	*Contributing CFR Staff*

FOREWORD

In the post–World War II period, the United States designed and championed an international economic system optimized for growth and efficiency through greater trade liberalization and global integration. That rules-based system, and the benign international economic environment it fostered, is currently subject to stress. Some of the stress is due to traditional protectionism, but it also reflects a major new development—the emergence of economic security as an increasingly central organizing principle for international economic policy.

Economic security, the convergence of national security and economics, is the new mantra. Over the past decade, the United States has increasingly intervened in the economy by deploying a range of offensive and defensive tools—chief among them tariffs, as well as export controls, restrictions on inward and outbound foreign investment, and industrial policy—to protect foundational technologies and resources deemed critical to the country's security. At the same time, the tenets of neoliberal economics have given way to harsh geopolitical realities, and power politics are rewiring the global economy. The contours of the new economic security paradigm remain to be fully defined.

Any administration—current or future—will be called on to determine under what circumstances the government should intervene in the market, how and when the government should deploy various tools at its disposal, and what limiting principles and guardrails should guide the instances and extent of intervention. Even as those issues are addressed, the United States is engaged in an unprecedented competition with China across military, political, economic, and technological domains. It is therefore imperative that we address the most immediate challenges facing U.S. economic security.

To help policymakers address those challenges, the Council on Foreign Relations convened a bipartisan Task Force on Economic Security under the auspices of its RealEcon: Reimagining American Economic Leadership initiative.

From the quantity and quality of artificial intelligence publications and patents to the production of drug inputs, China is well on its way to compete in, if not lead, the development of foundational technologies, including artificial intelligence, quantum (sensing, communication, and computation), and biotechnology. China has also discovered where it has leverage over the U.S. economy, including its near monopoly on critical minerals extraction and refining. Given the size and reach of the U.S. economy, export controls, investment restrictions, and industrial policy play an essential role in ensuring the United States maintains the technology edge over China, whether it is to correct for market failures at home or to reduce overdependencies on certain markets for critical inputs, such as advanced semiconductor chips.

Recognizing crucial challenges to investment, supply chains, and controls, the Task Force supports continued U.S. government leadership on the following specific recommendations: onshoring the manufacturing of critical inputs and components for semiconductors; accelerating U.S. progress on the world's first utility-scale quantum computer; creating a network of advanced biomanufacturing hubs with co-investment from the private sector; expanding the National Defense Stockpile to include critical minerals; developing the U.S. workforce by supporting the machinists, electricians, and other trades workers needed to advance those technologies; and establishing an Economic Security Center at the Department of Commerce.

At the end of the Cold War in the 1990s, the Council helped develop the concept of geoeconomics, the convergence of geopolitics and international economic policy. The current moment reflects an inflection point in the ongoing evolution of the U.S. role in the international economic environment not unlike thirty years ago. Thus, I commend the Task Force for providing policymakers with a pragmatic way to inform the United States' approach to economic security, grounded in deep knowledge and real-world experience.

I thank the cochairs, former Secretary of Commerce Gina M. Raimondo, former Deputy Secretary of the Treasury Justin G. Muzinich, and Chairman, President, and CEO of Lockheed Martin James D. Taiclet, for their leadership. I also thank CFR's Jonathan E. Hillman, who directed the Task Force and authored the report, and Anya Schmemann, who guided the process.

Michael Froman
President
Council on Foreign Relations
November 2025

EXECUTIVE SUMMARY

Strategic competition over the world's next generation of foundational technologies is underway, and U.S. advantages in artificial intelligence (AI), quantum, and biotechnology are increasingly contested. Economic security tools can help the United States win this competition and address several pressing risks, especially overconcentration of critical supply chains in countries of concern and underinvestment in strategically important areas. This report provides a comprehensive view of the vulnerabilities that the United States should address and offers practical recommendations for strengthening trust in supply chains and mobilizing the investment needed to prevail in these three crucial sectors of the future.

Challenges to U.S. technological leadership include the following:

- **China:** The Chinese government is spending heavily on AI, quantum, and biotech ($900 billion over past decade); making rapid advances in AI model performance, quantum communications, and biotech innovation; working to indigenize tech and dominate key sectors; and willing to weaponize chokepoints.

- **Investment:** Private capital avoids quantum and biotech due to long time horizons, lack of commercial demand, and scaling challenges; early financing for U.S. biotech start-ups dropped 65 percent in the first half of 2025; and China is spending twice as much as the United States on quantum.

- **Supply Chains:** The United States is dependent on China for rare earths (70 percent overall, 99 percent for heavy rare earths), data center and chip components (30 percent of printed circuit boards [PCBs], 60 percent of chemicals), biotech inputs and drug development (80 percent of key starting materials [KSMs], 33 percent of global active pharmaceutical ingredient [API] capacity, 80 percent of U.S. biotech companies have at least one Chinese contract), and single suppliers for quantum equipment (laser diodes, mirrors, amplifiers).

- **Controls:** Effective enforcement and monitoring of U.S. controls on foundational technologies requires tailored and efficient government capacity, technical expertise, and close partnership with the private sector.

The following recommendations would advance American leadership in tomorrow's technologies and reduce the leverage that other countries, especially China, could exploit during a conflict if tensions were to escalate:

- Build on the Trump administration's AI Action Plan by onshoring the manufacturing of critical inputs and components for semiconductors, including chemicals, printed circuit boards, and integrated circuit (IC) substrates.
- Accelerate development of the world's first utility-scale quantum computer through Department of Defense procurement, stimulating the U.S. private sector to meet the department's need.
- Establish a national network of advanced biomanufacturing hubs with private-sector co-investment and fund U.S. companies to build six-month stockpiles of KSMs and APIs from trusted markets.
- Secure critical minerals by expanding the National Defense Stockpile (NDS), accelerating permitting, and working with partners and allies to map sources and to pioneer recovery and substitution technologies.
- Build on the Trump administration's America's Talent Strategy, including by supporting the machinists, electricians, and other trades workers who are essential for leading in key technologies.
- Establish an Economic Security Center at the Department of Commerce that strengthens government coordination, technical expertise, and partnership with the private sector.

These targeted government actions, and others detailed in the report, are intended to unleash the U.S. innovation ecosystem, allowing the private sector to scale and diffuse technology globally and responsibly. Given the rapid pace of technological change, the institutional improvements recommended above would also position the United States to better respond to future changes and technologies that have yet to emerge. Looking beyond today's immediate challenges, the Task Force report concludes by offering principles to help U.S. policymakers decide whether and how to intervene in markets in the name of national security.

INTRODUCTION

The Rise of Economic Security

Economic power has long been an important foundation for national security and an enabler of U.S. military and diplomatic power.[1] In recent years, however, a series of global shocks have pushed economic power further to the front lines of national security policy. The COVID-19 pandemic disrupted global supply chains and exposed the downsides of concentrated economic interdependence. Russia's invasion of Ukraine shook global energy and food markets and exposed Europe's dangerous dependency on Russian energy supplies. China's actions—especially its massive subsidies aimed at dominating the commanding heights of technology and attempts to dominate critical supply chains—directly threaten U.S. economic growth and technological leadership, as well as the interests of U.S. partners and allies.[2]

Increasingly, economics and national security have converged, if not collided.[3] Around the world, policymakers are directly confronting a range of national security risks by turning to industrial policy, investment and export restrictions, and other economic tools (see figure 1). As policymakers embrace what many are calling economic security, they are increasingly grappling with the costs and trade-offs required in pursuing growth and competitiveness on one hand and resilience and national security on the other.[4]

In its broadest sense, economic security is anything related to the economy that affects national security. That includes, for example, research and development (R&D), the defense industrial base, and essentially any government policy that significantly shapes the contribution of the economy to national security. Since the U.S. economy underpins diplomatic and military power, an overly broad approach to economic security can quickly become impractical, as most economic activities touch on national security in some way.[5]

Government intervention in the economy in the name of national security is most clearly warranted in cases of market failure. Today, the market failures that loom largest for U.S. national security are shortfalls of private capital in strategically important areas and overconcentration of critical supply chains in unfriendly countries. The U.S. government also has an essential role to

FIGURE 1

Interventions in the Global Economy Are on the Rise

Number of restrictive measures announced globally

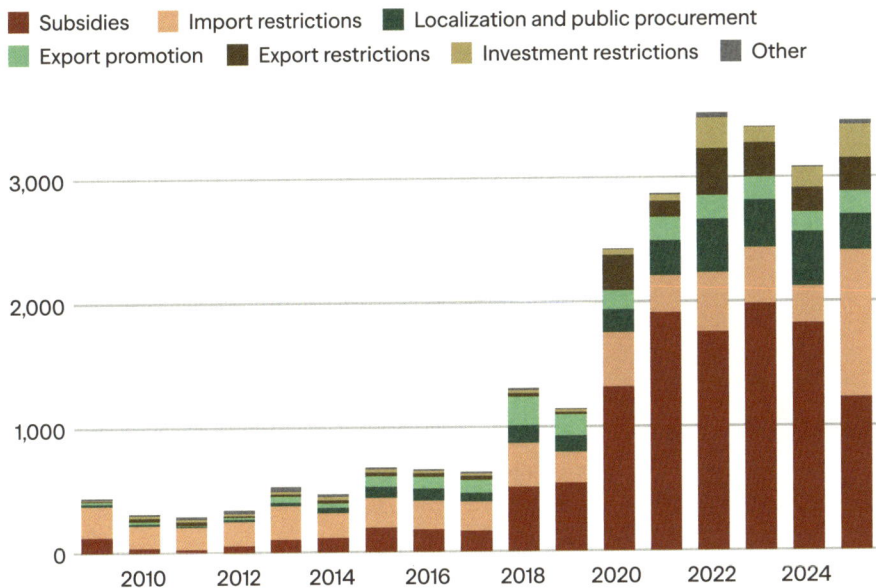

Note: Data for 2025 is as of September.

Source: Global Trade Alert

play in limiting the unintended spread of dual-use technologies and mitigating the risk of adversaries using such technologies. When the government is compelled to intervene, clearly communicating the national security risk and objectives of the intervention can help markets adjust accordingly.

To be sure, those developments are not entirely new, nor do they represent a break with traditional economics. Economists have long cautioned that markets can struggle to solve collective action and common goods problems, including the provision of national security. When national security needs to come first, there will naturally be costs to bear. The question then becomes one of timing. Should the United States bear the costs to mitigate risks now, or be prepared to pay a higher cost later if tensions escalate and it has less flexibility? Acting decisively in a targeted manner, as this report recommends in several areas, can strengthen security and ultimately reduce costs.

The Race for Tomorrow's Technologies

For the purposes of this report, the primary objective of U.S. economic security is to ensure American leadership in foundational technologies. Specifically, to generate actionable recommendations, the Task Force focused on how to ensure U.S. advantages in artificial intelligence (AI), quantum (sensing, communication, and computation), and biotechnology. This report's primary contributions are twofold: first, providing a bird's-eye view of vulnerabilities—which can sometimes be difficult for the U.S. government to spot, given siloing and bureaucratic challenges—and second, offering practical recommendations for mobilizing the investment needed to prevail.

Winning in these three areas of technology would contribute substantially to U.S. economic and national security, allowing the United States to take first and full advantage of the commercial, military, and other related benefits, such as shaping global rules and standards. Falling behind—especially to China—would allow others to gain those advantages and shape the international system to reflect their own interests.[6] Technological leadership can also help ensure that democracies remain resilient and free, safeguarding them from dangerous dependencies that authoritarian regimes could weaponize.

Each of these foundational technologies has vast dual-use potential (see figure 2). AI can expand access to information, accelerate scientific breakthroughs, and boost productivity, but it can also sharpen weapons for surveillance, misinformation, cyberattacks, and autonomous warfare. Quantum technology could both secure and jeopardize cryptography that protects data, cryptocurrency, and communications. Biotechnology could cure diseases and simultaneously create deadly viruses.

Certainly, additional technologies deserve attention from policymakers for their own merits, as well as their linkages to AI, quantum, and biotech. Nuclear fusion could power data centers. Robotics could give agency to AI systems. But focusing on AI, quantum, and biotech has the potential to offer broader lessons for the practice of economic security, given that these technologies are at different stages of development. Even more important, it is increasingly necessary to consider them in tandem.

The convergence and combined power of these technologies will become even more important to U.S. national security in the coming years. Tomorrow's systems for detecting and defending against biological threats, for example, could leverage AI to interpret massive amounts of data from air and water sensors, quantum imaging and sensing to detect small changes, and portable CRISPR-based diagnostic tools to test pathogens on site. Tomorrow's drones

FIGURE 2

AI, Quantum, and Biotechnology Have Critical Dual-Use Applications

Selected examples

ARTIFICIAL INTELLIGENCE

APPLICATION	CIVILIAN USE	MILITARY USE
Autonomous systems and surveillance Enable real-time object detection, route planning, and facial recognition	Taxis, city safety, industrial drones	Target identification and intelligence, surveillance, and reconnaissance
Large language models Can assist with summarizing, comparing, and interpreting unstructured data (e.g., reports, comms)	Research, legal discovery, business intelligence	Battlefield situational awareness, open-source intelligence, risk prediction

QUANTUM

APPLICATION	CIVILIAN USE	MILITARY USE
Quantum cryptography and quantum key distribution (QKD) Enable secure communications	Protection of financial data, telecommunications, health records	Hardened command-and-control in contested environments
Quantum sensors for navigation and detection Enable ultra-precise inertial navigation and electromagnetic /gravitational field detection	Autonomous vehicle guidance, mineral exploration	Resilient navigation, submarine or stealth aircraft detection

BIOTECHNOLOGY

APPLICATION	CIVILIAN USE	MILITARY USE
Synthetic biology for drug and bioweapon development Can be used to create vaccines or novel pathogens	Lifesaving treatments, vaccines, anticancer therapies	Viruses, toxins, bioweapons
Bio-sensing and environmental monitoring Detects pathogens, pollutants, or toxins	Early warning for disease outbreaks, food safety	Battlefield detection of chem/bioweapons, bio-surveillance

INTEGRATION: AI, QUANTUM, AND BIOTECH

APPLICATION	CIVILIAN USE	MILITARY USE
AI and quantum for material discovery and communication disruption Discover and model new materials but also conduct cyberattacks and perception manipulation, posing an urgent risk to digital sovereignty and command integrity	Faster development of high-efficiency batteries or superconductors	Compromising secure comms and situational awareness by disrupting encrypted channels, GPS signals, and authentic command messages
AI-driven models for drug discovery and gene editing Predict protein structures, optimize CRISPR targets, and simulate clinical trial outcomes for new treatments and targeted biological threats	Rapid development of affordable treatments for cancer and rare genetic disorders and shorter clinical trial timelines	Potential to engineer enhanced soldier resilience (e.g., fatigue resistance, wound healing), or to design highly targeted biological threats that exploit adversary population genetics

could harness AI, quantum sensors, and living organisms to enable operations in places where vision and GPS fail. On tomorrow's battlefield, U.S. and allied military personnel could gain advantages in everything from intelligence and targeting to logistics and lifesaving medicines.

Leading in these areas would benefit American workers and companies and unlock shared prosperity in markets around the world. Collectively, AI, quantum, and biotech could create up to $29 trillion in annual economic value by 2040, according to estimates from McKinsey & Company.[7] American manufacturing stands to benefit from the investments that will be needed to produce more semiconductors, printed circuit boards (PCBs), and other components essential to U.S. leadership in AI and quantum, and even more fundamentally, to manufacture new drugs and materials that biotechnology will make possible in the coming years. If U.S. companies become the world's preferred providers for these technologies, they will export more, supporting more jobs at home while helping set global standards (see figure 3).

The next decade will prove decisive, and speed is critical because fundamental breakthroughs could effectively end one era and begin the next. Cybersecurity researchers, for example, warn of "Q-Day," the moment when someone builds a quantum computer that can crack the most widely used forms of encryption. Being even months behind an adversary with this advantage could result in significant damage as financial and communication systems are compromised along with intelligence and military movements. Other game-changing capabilities have already arrived, such as AI models that can drive cyberattacks and overwhelm current defenses. First movers in technology do not always become dominant, but history suggests they are more likely to establish enduring advantages if they are also fast scalers.

Innovation and geopolitical competition are redefining the relationship between the government and private sector. In stark contrast to the Cold War, many of the most critical new technologies are now being developed in the private sector with private financing and not by government-sponsored programs and projects. That alters the historical model of managing the dual use of new technologies, as the U.S. government is much more likely to be an adopter rather than an inventor of breakthrough technologies. In other respects, the U.S. government is becoming more active and intervening in the market in new ways, such as taking equity stakes in private companies and sharing revenue from the export of strategically important goods.

There are certain risks to U.S. technological leadership that only the government can address, especially given that U.S. competitors are using nonmarket practices to their advantage. Competitors will not easily cede their control

FIGURE 3

Innovation to Diffusion: What Tech Leadership Looks Like

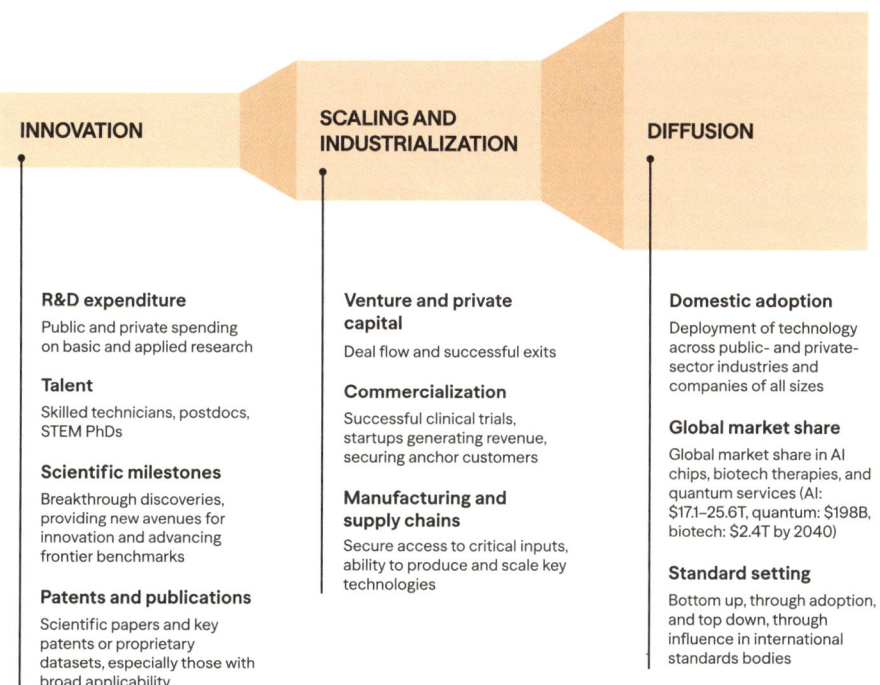

Source: CFR research

of critical minerals, ingredients for drugs, and other chokepoints—central nodes in the global economy where a dominant actor controls access and few substitutes exist.[8] Investment shortfalls also call for targeted government action, particularly for biotech and quantum technologies, which struggle to attract sufficient support from the private sector, given long time horizons, high technical risks, and significant capital requirements for development (see figure 4). Likewise, as the Trump administration's America's Talent Strategy recognizes, the government has an indispensable role to play in cultivating the human capital to compete in these areas by supporting education and training.

Smart government policy can contribute to the United States establishing a definitive lead in three technologies that will be essential to the future. Recognizing that the U.S. innovation ecosystem is multifaceted, this report aims

FIGURE 4

Public Investment Sparks Private Investment

Projected resulting private investment per **dollar of public investment** for select Task Force recommendations

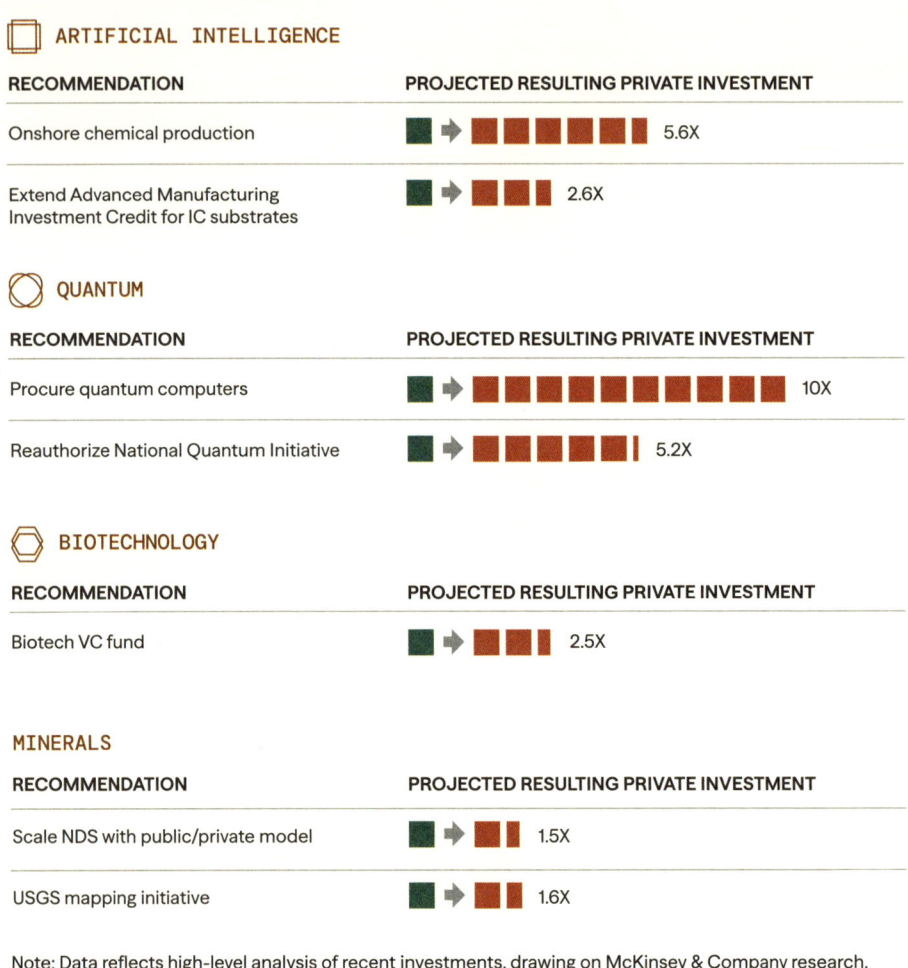

ARTIFICIAL INTELLIGENCE

RECOMMENDATION	PROJECTED RESULTING PRIVATE INVESTMENT
Onshore chemical production	5.6X
Extend Advanced Manufacturing Investment Credit for IC substrates	2.6X

QUANTUM

RECOMMENDATION	PROJECTED RESULTING PRIVATE INVESTMENT
Procure quantum computers	10X
Reauthorize National Quantum Initiative	5.2X

BIOTECHNOLOGY

RECOMMENDATION	PROJECTED RESULTING PRIVATE INVESTMENT
Biotech VC fund	2.5X

MINERALS

RECOMMENDATION	PROJECTED RESULTING PRIVATE INVESTMENT
Scale NDS with public/private model	1.5X
USGS mapping initiative	1.6X

Note: Data reflects high-level analysis of recent investments, drawing on McKinsey & Company research.
Source: McKinsey & Company

to make a practical contribution by focusing on how economic security tools can address shortfalls in investment, de-risk supply chains, and design and enforce effective controls that could otherwise jeopardize or undermine U.S. technology leadership.

China's Visible Hand

The greatest challenge to U.S. technological leadership comes from the Chinese government, which provides much greater support to strategic industries and has a higher tolerance for accepting failures alongside its successful bets. Over the past ten years, the Chinese government has spent an estimated $900 billion on AI, quantum, and biotech—more than three times U.S. government support for those technologies during the same period.[9] At home, Beijing favors Chinese firms and squeezes out U.S. and other foreign competition. China is not merely trying to tilt the playing field in its favor. It is playing a different game altogether.[10]

Chinese leader Xi Jinping's signature initiatives are designed to seize the commanding heights of technology and dominate foreign markets. Officials in Beijing are considering drafting a successor to their Made in China 2025 initiative, a plan launched in 2015 that aimed to dominate strategic sectors in a decade. Assessing the plan's performance, researchers at Bloomberg identified thirteen technology areas and concluded that Beijing has largely succeeded in five of them—including unmanned aerial vehicles, liquid natural gas carriers, and batteries—and has made significant progress in nearly all the others.[11] Chinese tech firms are active in over 165 markets globally and, over the past decade, have benefited from China's Digital Silk Road initiative.[12]

Xi understands the stakes, as he told an audience of China's top scientists and engineers last year:

> Cutting-edge technologies such as artificial intelligence, quantum technology, and biotechnology have emerged. . . . At the same time, the world is experiencing accelerated changes unseen in a century, with the technological revolution intertwined with great power competition. Increasingly, high-tech fields have become the forefront and main battleground of international competition. . . . There is an urgent need to further enhance the sense of urgency, intensify efforts in scientific and technological innovation, and seize the strategic heights of technological competition.[13]

China's playbook is producing results in these three industries of the future.[14] Chinese researchers lead globally in the quantity of AI publications and patents, and in 2024 alone, Chinese AI models reduced the gap with U.S. models in four key performance benchmarks by an average of 80 percent.[15] China has surpassed the United States in deploying quantum communications systems—it has launched the world's only quantum communication satellites and has established a national quantum communication network spanning over ten thousand kilometers.[16] China dominates the production of drug inputs—including those needed for antibiotics, fever reducers, and blood pressure medications—and accounts for nearly a quarter of all innovative drug development. In March, the Chinese government announced a new national venture fund that aims to invest an additional $138 billion over two decades into AI, quantum, and biotechnology.[17]

If China dominates these foundational technologies, the future will look very different. With $29 trillion at stake through 2040, Chinese companies stand to capture a windfall that would be measured in terms of economic growth, jobs, and additional support for R&D, which could ultimately cement China's advantages in technology for decades beyond. U.S. businesses, meanwhile, could find themselves caught in a vicious cycle, as they struggle to compete in foreign markets with less advanced products and services and struggle to innovate at home with less revenue to invest in R&D.

Chinese leaders have already previewed the perils of this scenario by demonstrating a willingness to weaponize China's near monopoly on critical minerals extraction and refining. They first cut off access to rare earths in 2010 in an attempt to coerce Japan, which was suddenly unable to obtain minerals essential for electronics and defense systems, among other technologies. Earlier this year, Chinese controls on rare-earth exports temporarily halted foreign manufacturing lines and threatened U.S. companies and workers in the automotive, technology, and defense sectors, before China lifted the restrictions as part of trade negotiations. If China controls the supply chains underpinning tomorrow's foundational technologies, from raw materials to advanced manufacturing, it will gain even more chokepoints that could be weaponized.[18]

In this scenario, the world would be more perilous for individuals and any government unwilling to bend to Beijing and safer for many authoritarian regimes. State censorship and mass surveillance could spread with Chinese products and standards. The proliferation of Chinese AI into more of the world's devices, essential services, and critical infrastructure would increase the risks of mass espionage and sabotage, with potential damage magnitudes worse than recent cyber operations like the Salt Typhoon hacks.[19]

Decisive military advantages hang in the balance. If the People's Liberation Army (PLA) gains superior AI, its forces could anticipate enemy movements, optimize logistics, and dominate the battlefield by deploying drones and other autonomous systems with unmatched speed and efficiency. Quantum superiority could enable the PLA to decode U.S. and allied communications and intelligence, while making its own communications impenetrable. The PLA could harness biotechnology to field lethal weapons that target individuals or populations based on genetic characteristics.

With these stakes in mind, the next section considers how to use economic security tools to advance U.S. leadership in AI, quantum, and biotech (see figure 5). The report concludes by recommending ways to upgrade U.S. economic security tools and offering principles to guide their use.

FIGURE 5

Selected Tools From the U.S. Government's Economic Security Toolkit

TOOL	PURPOSE	U.S. GOVERNMENT AGENCY
Export controls and foreign military sales*	Prevent unauthorized transfer of sensitive technologies, strengthen alliances and defense supply chains.	Export controls: Commerce (BIS) enforces Export Administration Regulations (EARs) and licensing, Justice prosecutes and enforces penalties. Foreign military sales: Defense Security Cooperation Agency administers; State approves; Defense manages acquisitions. Congress reviews major cases.
Industrial policy*	Promote growth, resilience, and strategic industries through targeted investment.	Congress funds via legislation; White House, Council of Economic Advisors, and Treasury shape and operationalize policy.
Investment screening*	Safeguard infrastructure and technology from foreign influence via capital investment and manage risks from U.S. capital outflows to adversaries.	Treasury (CFIUS) reviews inbound transactions; Treasury, Commerce, State lead outbound screening; Defense, State, Commerce, Energy, Homeland Security contribute to inbound.
Sanctions	Restrict trade, financial transactions, or economic activities abroad to deter adversarial behavior and mitigate threats.	Treasury (Office of Foreign Assets Control) issues regulations and enforces compliance via civil penalties; Justice Department prosecutes violations.
Tariffs	Adjust imports to protect national security through Section 232 tariffs.	Commerce investigates threat and reports finding/recommendation to president, who decides and implements action.

*Focus of report

WINNING THE RACE

This section considers how to use economic security tools to advance U.S. leadership in AI, quantum, and biotech over the next five years. For each foundational technology, it summarizes the current state of play, identifies key challenges related to investment, supply chains, and controls, and recommends targeted government interventions. To be sure, leading in these areas will require using an even broader array of supporting government policies and progress in dimensions beyond the scope of this study. But economic security tools have a major role to play (see figure 6).

Artificial Intelligence

As the Trump administration's AI Action Plan observes, "The AI race is America's to win."[20] Winning in AI over the next five years means leading across the AI tech stack, capturing a dominant share of the global market for AI, and slowing the diffusion of advanced capabilities to adversaries (see figure 7). Thanks to strong private investment, the United States leads globally in AI innovation and the buildout of AI data centers. But China is closing the performance gap, is taking a greater share of key data center components, and has a near monopoly on rare earths extraction and refining. To stay ahead, the U.S. government will need to make targeted investments that build manufacturing capacity and adopt security measures that strengthen export control enforcement and protect AI data centers.

INVESTMENT: U.S. PRIVATE SECTOR DRIVING AI INNOVATION

While the United States still develops the most advanced AI models, its lead is shrinking as others, mainly China, make headway. Last year, U.S.-based institutions produced forty notable AI models, according to the Stanford AI Index, as compared to China's fifteen. China's models are advancing, however, with major benchmarks approaching parity in performance, and the country has greater access to data and human capital that could provide an edge in the years ahead.[21] As commercial adoption grows globally, the private sector is increasingly driving AI innovation, producing 90 percent of notable AI models in 2024, up from 60 percent the year prior.[22]

FIGURE 6

How the U.S. Can Win: Challenges and Recommendations for Tech Leadership

▶ Recommendations with allied component
▶ Recommendations without allied component

ARTIFICIAL INTELLIGENCE
Read more on page 20

AREA	CHALLENGES	RECOMMENDATIONS
SUPPLY CHAIN	• Semiconductor supply chain risks include imports for 60 to 80 percent of chemicals and reliance on Taiwan for IC substrates and advanced chip imports • Data center risks include a reliance on China for networking equipment and 30 percent of printed circuit boards, as well as imports required for cooling technologies and construction materials • Critical minerals are imported in significant quantities; 70 percent of rare earths and 100 percent of heavy rare earths are sourced from China	▶ Onshore U.S. chemical supply chain for semiconductors by 2030 ▶ Expand AI server PCBA manufacturing in U.S. through grants to set up domestic facilities ▶ Build an allied ecosystem in Southeast Asia organized around PCB manufacturing through debt for manufacturing expansion ▶ Extend the Advanced Manufacturing Investment Credit to support foreign investment for critical chip inputs like IC substrates ▶ Scale the National Defense Stockpile to include materials critical to AI ▶ Complete mapping of domestic AI-relevant mineral reserves and accelerate permitting for mining and refining projects ▶ Establish recycling infrastructure and pursue advanced recovery methods for AI-relevant materials
CONTROLS	• Reach of U.S. export controls exceeds the government's ability to enforce them	▶ Upgrade Commerce's enforcement capabilities, including IT systems, technical expertise, and ability to acquire and test foreign tech ▶ Increase penalties on export control violations ▶ Launch an ICTS investigation of AI data center supply chains

 QUANTUM
Read more on page 27

AREA	CHALLENGES	RECOMMENDATIONS
INVESTMENT	• Private investment is inadequate to accelerate utility-scale quantum computing, which is necessary to solve more real-world problems	▶ Procure, through the DOD, one utility-scale quantum supercomputer and one hybrid system with a quantum processor and AI-enabled supercomputer to stimulate demand for scaled quantum systems ▶ Reauthorize the National Quantum Initiative with a focus on strengthening international partnerships

 **QUANTUM**
Read more on page 27

AREA	CHALLENGES	RECOMMENDATIONS
SUPPLY CHAIN	• Quantum control, components, and environment have several single suppliers (blue gallium laser diodes, high-electron transistor amplifiers, dilution fridges) • The United States is 100 percent import reliant for minerals, such as holmium-copper 2, strontium, tantalum, and indium	▶ Scale the National Defense Stockpile to include materials critical to quantum technologies ▶ Complete mapping of domestic quantum-relevant mineral reserves and accelerate permitting for mining and refining projects ▶ Invest in substitution technologies like metamaterials and quantum alloys ▶ Modify the implementation of the Florence Agreement to prioritize sourcing of quantum research components from trusted producers
CONTROLS	• Recent controls on leading-edge quantum computers and related equipment require monitoring as technology develops	▶ Upgrade Commerce's enforcement capabilities, including IT systems, technical expertise, and ability to acquire and test foreign tech

 BIOTECHNOLOGY
Read more on page 33

AREA	CHALLENGES	RECOMMENDATIONS
INVESTMENT	• Private investment avoids the most advanced biotech research due to scientific uncertainty and commercial risk, including segments with market failures driven by unclear or absent demand signals such as pandemic countermeasures, antibiotics, and some small-market products	▶ Offer advance market commitments to drive private investment to priority biotech areas like rare diseases with weaponization potential ▶ Develop a nationwide network of advanced biomanufacturing hubs with private-sector co-investment to support innovation and commercialization ▶ Establish a biotechnology investment fund to support high-priority areas of national security technology by covering a portion of preclinical costs
SUPPLY CHAIN	• United States depends on China and India for active pharmaceutical ingredients (APIs) and key starting materials (KSMs), and on China for drug manufacturing • United States has insufficient biomanufacturing capacity, and biomanufacturing facilities can cost over $2 billion	▶ Build an allied ecosystem for drug and API production to reduce reliance on Chinese CROs and CDMOs ▶ Partner with the private sector to expand stockpiles of APIs and essential medicines with key inputs (APIs, KSMs)
CONTROLS	• U.S. government lacks visibility into outbound investments in countries of concern	▶ Expand outbound screening to include reporting on biotech investments in countries of concern

Source: CFR research

The private sector has also made the United States the world's top source of investment for AI. Last year, U.S. investment in AI totaled $114 billion, driven overwhelmingly by private investment ($109.1 billion).[23] In comparison, China's investment totaled an estimated $98 billion, with public investment making up a much greater share ($56 billion).[24] By 2030, companies will invest almost $7 trillion on data center infrastructure globally, with more than 40 percent of that spending expected in the United States.[25]

Keeping the AI buildout on track will also require navigating bottlenecks in energy and permitting. Training frontier models and running large data centers require access to reliable, abundant, and affordable power. AI data centers are completed six to sixteen months faster in China than in the United States, due in large part to streamlined permitting and faster connections to the grid—pressing challenges that the Trump administration's AI Action Plan aims to address.[26]

Despite the massive private investment flowing into AI, talent remains a key constraint to the advancement of models, infrastructure, and adoption. There was an estimated 50 percent hiring gap for AI positions in 2024, and around 70 percent of workers now need a skills upgrade to compete in the workforce.[27] Those gaps extend to academia as well, with growth in computer science enrollments outstripping growth in computer science faculty over the past decade.[28] The gap between demand and supply of qualified AI talent is expected only to widen in the future. In semiconductors, for example, the industry could face a global shortfall of approximately four hundred thousand engineers by 2030, with North America alone projected to be short one hundred thousand semiconductor professionals.

Retaining the lead in AI also hinges on clearing several obstacles to domestic adoption. The United States leads the world in AI users, but China is gaining ground.[29] Small businesses and the federal government, for example, have yet to fully adopt AI at the scale of larger firms.[30] Barriers to adoption include cost, lack of training, and questions about return on investment. The federal government faces the challenge of integrating AI with legacy information technology (IT) systems, concerns about data privacy and security, and procurement processes. The Trump administration's AI Action Plan aims to improve national adoption, including by deploying advanced AI capabilities, use cases, and AI talent across federal agencies and improving interagency coordination on AI procurement by the federal government. For the private sector, it seeks to establish regulatory sandboxes and convene stakeholders in specific sectors, such as health care, energy, and agriculture.[31]

FIGURE 7

The AI Tech Stack

The key components to develop and deploy AI systems

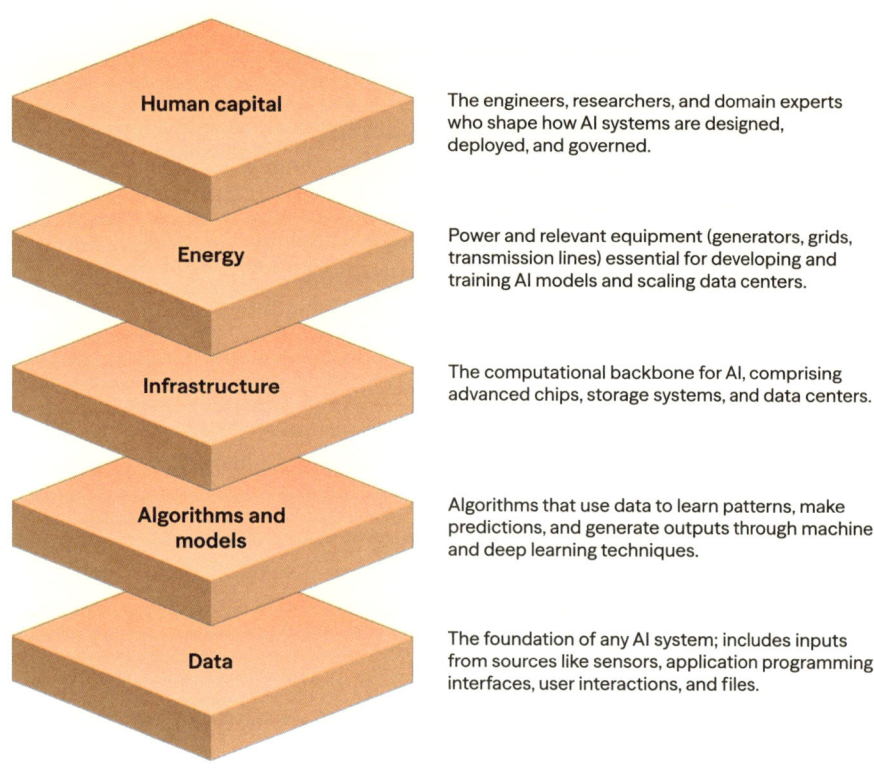

China is rapidly deploying AI with initiatives reaching into every corner of society. China's Ministry of Education has issued guidelines requiring AI to be woven into the national curriculum from elementary school through university. Beijing, Shenzhen, and other cities have launched "AI+" action plans through which the government promotes AI adoption, builds computing centers, and integrates AI into government services. Chinese state-owned banks have developed lending programs specifically for AI-related industries. Government directives have also accelerated the integration of AI with public services and industrial applications, with guidelines aiming for a 90 percent AI adoption rate across the entire Chinese economy by 2030.[32] The Chinese government's strong support and focus on practical applications, rather than artificial general intelligence, could provide an edge in producing low-cost solutions to sell globally.

SUPPLY CHAIN: DEPENDENCIES IN SEMICONDUCTORS, DATA CENTERS, AND CRITICAL MINERALS

With substantial capital supporting AI development and deployment in the United States, the primary near-term risks to U.S. leadership stem from supply chains. The U.S. semiconductor industry is expected to produce 23 percent of the world's leading-edge chips by 2030, up from 15 percent in 2024, following $450 billion in private investment sparked by recent manufacturing incentives.[33] But as Table 1 illustrates (see Appendix), the United States still faces several chokepoints for AI in semiconductors, data center components, and critical minerals.[34] To be sure, components sourced from allies and partners present a lower level of geopolitical risk than those sourced from adversaries. However, this analysis also identifies components that are highly concentrated in small numbers of friendly countries. As the COVID pandemic demonstrated, supply chains risks are not limited to foes.

Several semiconductor supply chain risks stand out. Chip fabrication requires wet chemicals and dry etchants that are essential for cleaning and etching patterns on silicon wafers. China is a key source of those inputs, while Japan produces silicon wafers and photoresists that transfer circuit patterns onto wafers. Physical dependencies include integrated circuit (IC) substrates, used to connect chips to PCBs, from Taiwan, South Korea, and Japan and advanced manufacturing equipment sourced from the Netherlands and Japan. Taiwan plays an outsized role by hosting much of the world's leading-edge research and design capabilities; advanced packaging capacity, which stacks and integrates chips after fabrication; and upstream inputs for packaging and insulation, such as polyethylene and polypropylene resins.

Chinese government subsidies have undercut the ability of the United States and partner countries to compete in the market for legacy and power semiconductors. Legacy chips are less advanced but remain important, given their use in cars, consumer electronics, and defense products. Power chips are used for managing high voltages and currents. Thus far, U.S. policymakers have responded with tariff investigations into semiconductors, manufacturing equipment, and legacy nodes. Congress has also considered a range of restrictions to prevent U.S. government agencies from procuring semiconductor products and services from China.[35] But these measures still leave a range of risks related to other parts of the AI supply chain, particularly for data centers (see figure 8).

FIGURE 8

Selected Supply Chain Risks for U.S. AI Data Centers

A **Construction materials (steel, copper, aluminum)**
Enables infrastructure build out
Imported in significant quantities

B **Networking equipment**
Facilitates base data center operations
Main supplier: China

C **Cooling technologies**
Manage heat and energy consumption of data centers
Main suppliers: Taiwan, Mexico, Germany

D **Printed circuit boards**
Support key operations like data processing, storage, and network communication
Main supplier: China

E **Logic chips**
Used for processing, execution, and inference
Main supplier: China

F **Chemicals**
Used for wafer manufacturing, cleaning, and chip etching
Main supplier: China

G **Diesel backup generators**
Provide an alternative power source
Main suppliers: UK, Brazil, UAE

Not shown **Other power equipment**
Ensures smooth transmission and distribution of power from grid to data centers
Main suppliers: China, Mexico

Source: CFR research

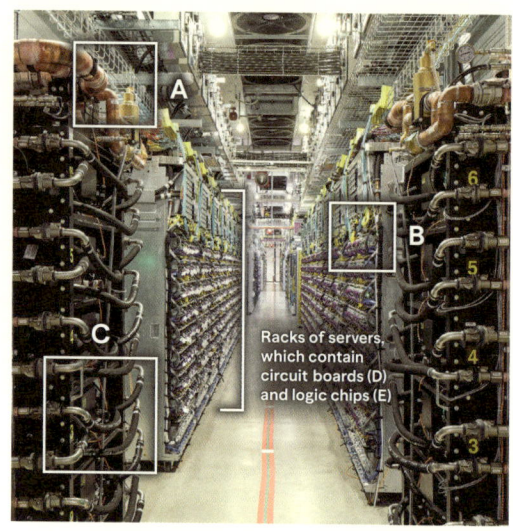
A Google data center in Ohio
Google

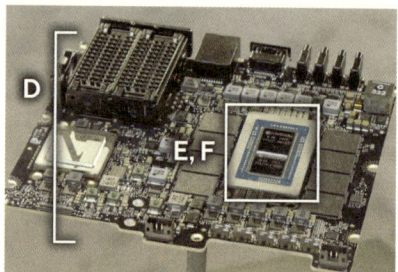
A Nvidia DGX Spark Chip/Board GB 10
Chesnot/Getty Images

A Google data center in Texas
Google

For data centers, supply chain risks include key power equipment and components—such as gas turbines, transformers, and uninterruptible power supplies—that are produced domestically but whose inputs are imported and have long lead times due to high demand. The production of battery energy storage systems (BESS), which can stabilize power grids and facilitate data center deployment in rural areas, relies heavily on imports.[36] Additionally, PCB design and manufacturing and PCB assembly (PCBA) manufacturing take place primarily in China and Taiwan, and the cost differential remains too high for companies to justify moving to the United States. The United States must also remain vigilant in areas where it has a lead, such as next-generation cooling and advanced optical transceivers, the latter of which is critical for transferring information within data centers. From 2017 to 2023, Chinese government incentives helped Chinese firms gain market share of advanced optical receivers largely at the expense of U.S. providers, whose market share fell from 67 percent to 43 percent.[37]

China's near monopoly on critical minerals extraction and refining is another significant risk, given the importance of those inputs for semiconductor manufacturing and data center components. The United States relies on China for 70 percent of its rare earths and nearly 100 percent of its heavy rare earths, which together are used for polishing semiconductor wafers and as insulation for advanced chips, among other applications.[38] The United States is completely dependent on China for all of its arsenic and holmium copper, which are critical for producing silicon chips and quantum cryocoolers, respectively. Though other countries, including the United States, boast significant critical mineral reserves, they do not have scaled infrastructure to refine mineral concentrates into usable compounds. China's vast refining and processing capacity, therefore, poses an additional challenge to de-risking efforts, as minerals mined elsewhere often pass through China.

CONTROLS: REACH EXCEEDS GRASP

The United States has struggled to enforce export controls on advanced semiconductors. Since 2022, the primary goal of these controls has been to maintain as large a lead as possible over competitors. In practice, however, advanced semiconductors have continued flowing to prohibited countries, underscoring the basic challenge of controlling goods in a global economy and the importance of persuading other nations to adopt and enforce the same controls.[39] China remains one to two generations behind in semiconductor production but has made progress in deep ultraviolet lithography and advanced packaging technologies. As these trends highlight, controls can at best slow an adversary, but only temporarily. That is why they must form part of a longer-term strategy that keeps U.S. technology ahead in performance and adoption.

Recent experience also highlights challenges in the U.S. government's export control regime. The Commerce Department's Bureau of Industry and Security (BIS) struggles to attract technical expertise on par with industry to design controls and relies on antiquated data systems to monitor them. BIS's staff capacity has not kept pace with the growth in controls. In 2021, there were more than 32 million U.S. exports of dual-use items, and BIS had only 190 enforcement agents and analysts, including only three agents in China and Hong Kong.[40] Finally, the penalties for violating export controls have remained relatively mild, and, because they rarely approach the value of the illegal transactions, they do not sufficiently deter violators. The largest export control fine to date, $300 million, was more than twice the estimated profit of the illicit transaction but significantly less than its revenue.[41]

RECOMMENDATIONS

The Trump administration's AI Action Plan includes recommendations to accelerate innovation, build infrastructure, and lead internationally.[42] Building on those actions, the following recommendations focus on addressing supply chain vulnerabilities and strengthening controls. Recommendations for developing the U.S. workforce and improving access to critical minerals are included near the end of the "Winning the Race" section, given their broader relevance. To that end, the U.S. government should consider the following actions:

Support American manufacturing and trusted sourcing from partners and allies by offering incentives to shift supply chains away from unfriendly countries. Without targeted incentives, supply chains will remain overly concentrated. Analysis of recent investments by McKinsey & Company suggests that leveraging $4.5 billion in public incentives and foreign investment could catalyze an additional $20 billion in private investment and directly support up to 37,000 jobs in the priority areas below. Building and operating these manufacturing facilities will also depend on ensuring adequate access to supporting infrastructure, such as water and water treatment, grid connectivity and power generation, and transport and logistics—requirements that present opportunities for additional growth and jobs.

- **Chemicals:** The U.S. government should incentivize companies to onshore the production of chemicals for semiconductors by 2030, including ultrapure wet chemicals, dry etchants, and photoresists. Japan is a natural partner for this effort, given that the U.S.-Japan trade and investment agreement includes $550 billion of Japanese investment into the United States, including for the semiconductor industry, and Japanese firms are leaders in relevant areas, such as photoresists. Approximately $3 billion would bridge the gap in startup costs between foreign facilities and those in the United States by covering up to 25 percent of the initial investment for up to 50 U.S. facilities, which would support 15,600 jobs. Those incentives would attract up to $17 billion in additional private capital by capping funding at $300 million for the most expensive facilities. (See Appendix for more details on chemical dependencies.)

- **Printed circuit boards:** The U.S. government should pursue a dual-track strategy to expand PCB and PCBA manufacturing capacity. Domestically, $900 million in grants could subsidize up to 35 percent of capital costs for 5 to 7 new dense AI server PCBA facilities, mobilizing $1.6 billion in private investment, supporting up to 8,200 jobs, and shifting 5 to 10 percent of global server PCBA production to the United States. Abroad, the U.S. International Development Finance Corporation (DFC) should provide some $2.5 billion in debt financing to support PCB manufacturing expansion by U.S. businesses in key markets, such as India and Malaysia, complementing Thailand's own successful incentives and recent progress.

- **IC substrates:** Congress should extend the Advanced Manufacturing Investment Credit beyond its 2026 sunset and expand it to support critical semiconductor inputs, particularly IC substrates, which serve as the base materials connecting chips to printed circuit boards. This incentive initially offered a 25 percent tax credit for a facility producing semiconductors or semiconductor manufacturing equipment. A targeted expansion of the investment credit would cost $750 million and could unlock over $2 billion in foreign direct investment from major IC substrate companies and support up to fourteen thousand jobs.

Strengthen controls and enforcement. Priorities for action include the following efforts:

- **Upgrade expertise, technology, and authorities at the Department of Commerce's Bureau of Industry and Security** by providing flexible hiring authorities for technical experts; adopting modern analytics and data capabilities to better design export controls, identify evasion, and evaluate effectiveness; and codifying authorities for the Office of Information and Communications Technology and Services (ICTS) while protecting free speech. Those controls would benefit AI export control enforcement most immediately, as well as position the U.S. government to better manage nascent controls in quantum, biotech, and other emerging technologies.

- **Increase penalties on export control violations** closer to the value of the illicit transaction and extend penalties to financial facilitators. Experience with sanctions shows that significant penalties incentivize compliance and deter violations, especially when extended to include the banking system, which is concentrated and has robust compliance programs. The Department of Commerce should utilize its authority to levy administration penalties that are up to twice the revenue of the illicit transaction. The Department of Justice and the Department of the Treasury should increase banks' liability for knowingly facilitating export control violations. Effective controls will need to affect the incentives of financial facilitators, not just the exporters and re-exporters.

- **Create a fast technology teardown program** among relevant U.S. government agencies, including the Department of Commerce, Department of Defense (DOD), and National Security Agency, to rapidly acquire and analyze hardware and software from foreign sources. That would allow the U.S. government to more quickly assess how foreign technology is progressing, including its capabilities, origins of components and source code, and effectiveness of any related U.S. and allied controls. This program would leverage special acquisition authorities at the Defense Department and the intelligence community, as well as their expertise and that of specialists at the Commerce Department.

- **Conclude an ICTS investigation of AI data center supply chains** to better assess risks and potential responses, including prohibitions on some foreign components, such as transceivers. BIS's ICTS office, which is leading this investigation, would also benefit from having Congress codify its authorities to investigate, mitigate, and prohibit ICTS transactions involving foreign adversaries. Codification would help ensure consistent application of these authorities and guard against legal challenges.

Quantum Technologies

Quantum technologies—including computing, communications, and sensing—are more nascent than AI. Quantum computing uses quantum bits, or qubits, to enable faster processing of complex calculations. Quantum communications use the principles of quantum mechanics to drastically improve the security of communications. Quantum sensing uses similar principles to allow for extremely precise measurements at the atomic level.

Winning in quantum over the next five years means leading in R&D and being the first to reach utility-scale quantum computing. Of particular importance are "gates-based" approaches, which are the most general purpose and theoretically powerful, as they can, in principle, run any quantum algorithm. The United States currently leads in quantum computing and sensing technologies, but China leads in certain aspects of quantum communications and is investing heavily in general quantum research and development. Given that there are several competing approaches to delivering the underlying technology, any one of which could prove viable, the U.S. government should focus on providing targeted support to reduce the cost of pilot facilities for testing equipment and on securing supply chains for the most cost-effective, scalable approaches. Since quantum efforts are already international in nature, the United States will need to adjust export controls to cover new approaches and equipment with allies and partners.

INVESTMENT: PUBLIC INVESTMENT NEEDED TO ACCELERATE UTILITY-SCALE COMPUTING

Computing represents the biggest economic opportunity of the three quantum technologies and could create $1 trillion to $2 trillion in economic value by 2035.[43] "Utility-scale" quantum computers that solve real-world problems that are impractical on classical computers are anticipated to arrive around 2030, although more conservative projections put the moment around 2035 to 2040. Regardless of when the breakthrough arrives, achieving it will require major investments in hardware. Estimates vary but often land in the neighborhood of several billion dollars to scale production of specialized components for quantum computers.

The path to scaling quantum computing is especially challenging for companies that lack revenue from other business lines, as well as larger public companies accountable to shareholders, due to high R&D expenses over long periods of time and capital costs for pilot facilities. The direct market for quantum computing is estimated to reach $28 billion to $72 billion by 2035, a wide range that reflects a high degree of uncertainty and risk for private investment.[44] A long time horizon, a lack of commercial demand, and technical uncertainties make it challenging for quantum-related start-ups and large public companies to secure capital, especially those developing hardware.

Government support for quantum research is growing faster outside of the United States. In 2018, Congress passed the National Quantum Initiative

(NQI), which provided a coordinated federal strategy and over $1.2 billion in funding for quantum information science, including research centers, workforce development, and interagency collaboration between the Department of Energy (DOE), the National Institute of Standards and Technology (NIST), and the National Science Foundation (NSF). But it partially lapsed in September 2023, leaving a gap in support that Congress is working to address through reauthorization. An encouraging example of international partnership is the collective investment by the Defense Advanced Research Projects Agency (DARPA), the state of Illinois, and the government of Australia in PsiQuantum, a U.S. start-up that is building quantum computers in Chicago and near Brisbane.[45] Other countries, including Germany, India, South Korea, and the United Kingdom, have all announced significant new funding streams in recent years, joining Canada, France, Israel, Japan, and the Netherlands, among others, in making quantum technologies a national priority.

Quantum science is a significant national priority for China, where government support extends across R&D, application domains, and deployment. As Xi told a session of the Political Bureau of the Chinese Communist Party in 2020, "Developing quantum science and technology is of great scientific and strategic significance."[46] As of 2023, China had provided $15.3 billion in public support for quantum, more than twice the combined U.S. public- and private-sector support.[47] The lion's share of that support, $10 billion, has gone toward building a ninety-one-acre National Laboratory for Quantum Information Sciences, which functions both as a research institute and as a national test bed for translating quantum science into military and commercial uses.

China leads the world in quantum communications and is making significant strides in quantum sensing and computing. It has constructed the world's largest quantum communications network, which spans over 10,000 km and integrates quantum key distribution, a technology that increases security of communications and allows detection of eavesdropping.[48] China has also launched two quantum satellites, enabling deployment of quantum communications to other countries, such as South Africa, in March 2025.[49] While China's sensing capabilities are limited to the laboratory scale, it is making progress and has advantages to leverage in research output and deployment.[50] In 2021, China built a fully domestic superconducting quantum computer that reportedly surpassed some U.S. systems at the time in speed and processing power but has since fallen behind in key performance measures, such as qubit count and error correction.[51]

Workforce development is an important challenge for the United States in this field, as demand for quantum talent is rapidly outpacing supply, with limited academic pathways, a small pool of trained specialists, and a growing gap between research and commercial skills. Despite being a global leader in quantum technologies, the United States is fourth in quantum talent availability behind the European Union, China, and India. In a poll by the Quantum Economic Development Consortium, a leading global industry group, 92 percent of its members agreed that there is a shortage of U.S. citizens and permanent residents with quantum qualifications.[52] With quantum computing projected to generate 250,000 jobs globally by 2030 and 840,000 by 2035, the United States risks watching those opportunities go elsewhere.[53]

SUPPLY CHAIN: COMPONENTS AND CRITICAL MINERALS

Table II (see Appendix) summarizes the main vulnerabilities for quantum technologies across four categories: control, components, environment, and critical minerals. Given that there are several plausible approaches to scaling quantum technologies, each with their own set of inputs, this supply chain analysis focuses on the most common components (see figure 9).

Reliance on China for precision lasers, holmium-copper 2, and commercial off-the-shelf components, like PCBs, nonlinear crystals, and mirrors, are among the top supply chain risks. Precision lasers are essential for several modalities, as they control qubits and laser-cool atoms, and Chinese lasers dominate the market due to their high quality and low price. Ninety percent of holmium-copper 2, a rare-earth alloy critical for cryocoolers needed for superconducting qubits, is imported from a single producer in China. China is also a major source of components used to control quantum systems, comprising 30 percent of U.S. PCB imports and 42 percent of global nonlinear crystal production.

The supply chains for quantum technologies also highlight the importance of working closely with allies and partners as part of a broader de-risking strategy. Finland is the primary source of specialized laser components (called semiconductor saturable absorber mirrors), as well as dilution refrigerators, which create the ultra-low temperatures needed for qubits to operate. Japan is the sole or primary producer of blue gallium nitride laser diodes and 200 mm sapphire wafers and can produce holmium-copper 2. Germany is a key source of ultra-high vacuum chambers, laser diodes, superconducting nanowire single-photon detectors, rubidium-87, and strontium. Other allies and partners—including Canada, Mexico, South Korea, and the UK—either produce components, such as vacuum chambers and dilution fridges, in smaller quantities or supply minerals, like indium, strontium, and aluminum.

FIGURE 9

Selected Supply Chain Risks for U.S. Quantum Computers

A Cryocoolers and compressors
Provide the initial stages of cooling
Main suppliers: China, domestic, Finland

B Superconducting coaxial cables
Enable ultra-low-loss transmission signals
Main suppliers: Japan, UK, China

C Holmium-copper 2
Rare earth alloy critical to cryocoolers
Main supplier: China

D High electron mobility transistor amplifiers
Allow for fast signal processing and low noise
Main supplier: Sweden

E Helium-3
Produces cooling power in dilution refrigerator
Main suppliers: Domestic, Russia, China

F Dilution refrigerator
Creates ultra-low temperatures for superconducting qubits
Main suppliers: Finland, UK, domestic

Not shown Control electronics/ cryogenic CMOs
Manage quantum operations and manipulate qubits
Main suppliers: Israel, Australia

Source: CFR research

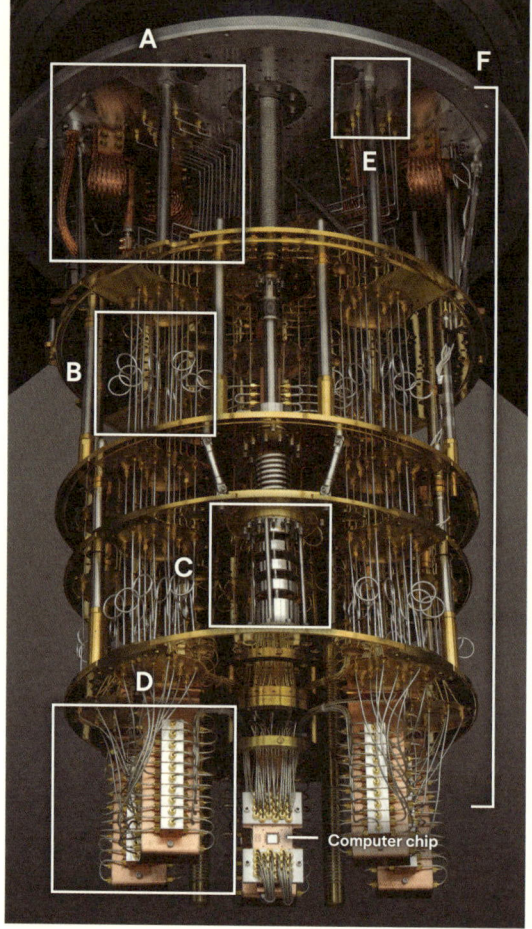

An IBM model of a quantum computer
James Estrin/New York Times/Redux

CONTROLS: NASCENT MEASURES REQUIRE MONITORING

In September 2024, the United States announced worldwide export controls on key equipment, materials, and software for quantum computers.[54] Going beyond earlier controls, these measures focus on leading-edge quantum computers and related equipment, preemptively capturing technology that does not yet exist. Importantly, these controls are aligned with key allies, including Australia, Canada, France, Germany, Italy, Japan, Spain, and the UK. Given the nascent state of quantum technologies and the relatively recent expansion of these controls, the near-term priority should be monitoring and assessing their effectiveness. Controlling software could prove challenging, given open-source development and release of these tools.

RECOMMENDATIONS

Procure two quantum computers. Congress should allocate $1.3 billion of initial funding to allow the DOD to procure one utility-scale quantum supercomputer and one hybrid supercomputer. Of the funding, $1 billion would support the first phase of procuring a utility-scale system. The DOD could leverage Other Transaction Authority—a flexible contracting mechanism that allows for engagement with nontraditional defense vendors—to work through established consortia. The contract should use milestones aligned with DARPA's Quantum Benchmarking Initiative, so that payments are contingent on making progress toward well-defined goals. Additional funding would likely be required as the technological modalities and system architectures for utility-scale quantum computing mature. The $300 million would support procuring a hybrid system that integrates a quantum processor with an AI-accelerated supercomputer, a high-performance system that combines traditional central processing units with AI chips. Those efforts could spur an additional $13 billion in private investment.

Reauthorize the National Quantum Initiative with an emphasis on deepening international partnerships for research and supply chain security. Since 2018, the NQI has established five Quantum Information Science (QIS) Research Centers (at the Department of Energy) and five Quantum Leap Challenge Institutes (at the National Science Foundation) and supported research into promising quantum approaches, materials, and programming models, among other areas.[55] It also paved the way for the United States to establish eleven bilateral partnerships for QIS research and supply chain security, including with Australia, Germany, Japan, and South Korea. The proposed NQI Reauthorization Act, which would expand the program's national infrastructure, extend its duration to December 2034, and authorize $2.7 billion over five years, was not brought for a vote in 2023 or 2024, though it was the subject of a hearing this year. Congress should renew these core provisions and build on the International Quantum Cooperation Strategy, previously proposed by the National Science and Technology Council's Subcommittee on Quantum Information Science in 2024, by deepening cooperation with allies and jointly assessing research and supply chain barriers.[56]

Increase restrictions on U.S. research institutions acquiring quantum equipment from foreign countries of concern.[57] The U.S. government should modify the implementation of the UN Educational, Scientific and Cultural Organization (UNESCO) Florence Agreement, a 1950 treaty that waives customs duties on scientific materials, to give preference to U.S.-manufactured research equipment for quantum and waive import duties for items purchased from allies and partner countries, such as quantum-enabling lasers, optics, and photonics components. In cases where there are sufficient alternative and affordable sources of supply, the United States should not grant these preferences to items from foreign countries of concern. Sourcing requirements for research institutions could be added to federal funding, including through reauthorization of the NQI, but should also take into account the availability of substitute equipment and strive to offer resources to offset cost increases. Additionally, the Commerce Department could pursue an ICTS investigation of security risks related to quantum components.

Biotechnology

Biotechnology has the potential to transform the physical world in the coming years by harnessing cellular and biomolecular processes. At stake is control of tomorrow's production methods for everything from lifesaving medicines to high-efficiency crops and livestock to chemicals and critical minerals. In the future, the military could harness synthetic organisms to protect soldiers, repair equipment, generate fuel, and produce water in challenging environments. As Senator Todd Young and Dr. Michelle Rozo, who lead the National Security Commission on Emerging Biotechnology, have explained, "Emerging biotechnology, coupled with artificial intelligence, will transform everything from the way we defend and build our nation to how we nourish and provide care for Americans."[58]

Winning in biotechnology means leading in R&D and global market share, limiting access to high-risk biological agents and tools, and securing supply chains for critical medicines and vaccines for the United States. China is closing the gap with the United States and now accounts for nearly a quarter of global clinical trials and early drug development.[59] U.S. industry has grown increasingly dependent on foreign (largely Chinese) suppliers for key inputs, as well as R&D. The U.S. government must reduce these dependencies while doubling down on innovation.

INNOVATION: CHINA GAINING GROUND

Over the next five years, the global biotechnology market is poised to reach $1.5 trillion to $2.2 trillion, with pharmaceuticals composing the largest share.[60] Home to several top biotech companies and technologies, the United States has historically benefited from a strong and open innovation ecosystem composed of public and private research institutions, funding for foundational R&D, and an established pipeline for drugs to go to market. The United States leads in several important areas of biotechnology—including genetic engineering, vaccine research, and agricultural technology—but China could soon take the lead.[61]

An emerging constraint to U.S. leadership in biotechnology is human capital. Nearly a fifth of the U.S. workforce in life sciences is aged fifty-five or older, posing risks of significant leadership and expertise loss through retirement in the coming years.[62] Replacing these workers and filling new positions will require overcoming bottlenecks in the talent pipeline. Fewer than 30 percent of public high school biology classes cover molecular biology, which is essential for biotech. Meanwhile, graduate programs are not keeping pace with industry needs, particularly in R&D, clinical research, and technical positions. For every experienced clinical research coordinator seeking work in the United States, for example, there are seven jobs posted.[63]

China's strengths are most evident in its dominant production of active pharmaceutical ingredients (APIs) and key starting materials (KSMs) for drugs, boasting its ability to produce high-quality inputs more cheaply and more quickly than anywhere else. It is the primary outsourcing destination for large U.S. pharma companies, which rely on contract R&D labs for drug discovery and research. Backed by massive state subsidies, favorable regulation, and vast human capital, Chinese firms can advance products quickly and cheaply, often licensing them back to U.S. companies or flooding the global market with low-cost "fast followers."

China is no longer just a manufacturing hub for biotechnology but is now a leading innovator of drugs, with its own innovation ecosystem accelerating. Its biopharma R&D spending has increased four-hundred-fold over the past decade, and its share of global biotech patents has surged from 1 percent in 2000 to 28 percent in 2019 (surpassing the U.S. share of 27 percent). Those trends are undercutting U.S. biotech investments and licensing activity. Last year, a third of global licensing deals involved Chinese-developed assets. Just this year, Pfizer and Merck struck $8 billion in landmark deals with Chinese biotech companies, demonstrating how U.S. pharma companies are aggressively sourcing high-potential assets in China.[64] The United States must compete in R&D and patents for the next generation of biotech, even as it works to address supply chain vulnerabilities, or it will be relegated to producing yesterday's tech as China captures the market for tomorrow's.

SUPPLY CHAIN: DRUG INPUTS AND MANUFACTURING

The United States relies on China for basic inputs to drugs, as well as for R&D and manufacturing services, as summarized in Table III (Appendix).

The numbers speak for themselves:

- China or India supply 60 percent of APIs for U.S. medicines; India supplies nearly half but is itself reliant on Chinese-sourced key starting materials.[65]

- Around 40 percent of finished drugs consumed in the United States are imported, and roughly 80 percent of KSMs for those drugs are sourced from China.[66]

- Of 102 drug products where a single supplier controls at least 75 percent of the market, 32 are sourced from China.

- Only 10 percent of generic U.S. APIs are produced domestically, and Chinese APIs are present in nearly 25 percent of drug volume sold in the United States.[67]

- No U.S. manufacturing source exists for more than 80 percent of the active ingredients in medicines the Food and Drug Administration deems essential for public health, and around 20 percent of these vital APIs appear to come only from China.[68]

Trends are troubling as well. Over the past five years, U.S. pharmaceutical companies have gone from not licensing any drugs from China to sourcing one-third of their new drug candidates from Chinese biotech firms.[69] If access to these inputs and drug candidates were lost, the United States would face shortages in antibiotics, fever reducers, and other important medicines (see figure 10). For example, 80 percent of the global supply of PAP, a precursor for acetaminophen (the API in common pain relievers), is sourced from China.[70]

FIGURE 10

Selected Supply Chain Risks for U.S. Biotech

Bioproduction and fermentation

Cultivation of biologics via microbial or cell culture in bioreactors

Construction of domestic facilities is costly and has long wait times

Genetic manufacturing

Tools and facilities for gene/cell therapy

Long wait time for good manufacturing practice (GMP)–grade viral vectors; GMP facility cost; limited contract manufacturing organization bandwidth

Active pharmaceutical ingredients (APIs)

Raw material input that delivers the intended therapeutic effect

Main suppliers: India, China

Key starting materials (KSMs)

Structural building blocks for APIs to synthesize

Main supplier: India, India depends on China

R&D outsourcing (CROs & CDMOs)

Contracted research organizations (CROs) and contract development and manufacturing organizations (CDMOs)

Most of global CDMO capacity in Asia and Europe; most companies tied to Chinese CDMOs

Source: CFR research

A bioreactor, used in bioproduction and fermentation

Production of a COVID-19 vaccine in China
Chen Xiaogen/VCG via Getty Images

At a time when the United States and China are moving apart in several areas of technology—such as semiconductors, cloud computing, and next-generation wireless networks—trends in biotechnology are heading in the opposite direction. U.S. companies are increasingly outsourcing a range of R&D, development, and manufacturing services—from preclinical trials to formulating and producing biologics—to companies in China that offer low costs and high quality due to large talent pools and government support. Despite risks to intellectual property (IP), incentives for outsourcing to these companies, known as contract research organizations (CROs) and contract development and manufacturing organizations (CDMOs), remain powerful. Smaller U.S. firms are especially dependent on outsourcing, as they lack the capital to maintain large R&D teams and manufacturing facilities.[71]

Nearly 80 percent of U.S. biotech companies have at least one contract or product agreement with Chinese firms.[72] U.S. companies began outsourcing API and generic drug production in the late 1970s, and China has rapidly expanded its role in these supply chains since joining the World Trade Organization in 2001. A single Chinese firm, WuXi AppTech, has been involved in developing one-fourth of the drugs used in the United States.[73] As a result, IP and production know-how is increasingly contained within a single Chinese company that has extensive ties to the Chinese government.[74] This arrangement creates a one-sided risk: the United States can make very little, and China knows how to make everything.

U.S. domestic biomanufacturing efforts are constrained by long lead times for key materials and equipment and high infrastructure setup costs. Under current quality standards, viral vectors that are used to deliver genetic material into cells for gene therapy or vaccine trials can take twelve to fifteen months to acquire. Bioreactors, which are essential for cultivating cells and producing biological products at scale, are another key bottleneck. New biomanufacturing factories can require up to $2 billion and take two to five years to build and operationalize.[75] In the absence of sufficient U.S. biomanufacturing capacity, companies will continue looking to Europe and Asia, which collectively account for roughly 80 percent of the global CDMO capacity.

CONTROLS: OUTBOUND INVESTMENT NOT COVERED

The United States and its allies have taken recent steps to strengthen controls on biotechnology. In January 2025, the United States announced new export controls on biotechnology equipment and related technology, including high-parameter flow cytometers and certain mass spectrometry equipment, both of which are used to generate large and detailed biological datasets. Those controls are primarily intended to restrict China's access to high-quality data for military purposes and are aligned with dual-use controls imposed by Australia, the EU, and the UK. In 2018, Congress expanded the jurisdiction of the Committee on Foreign Investment in the United States (CFIUS) to cover noncontrolling foreign investment in biotechnology and life sciences sectors, including licensing and real estate ties to biotech firms managing sensitive genetic data. Biotechnology is not yet included in the U.S. outbound investment screening regime, which could limit the U.S. government's visibility into foreign transactions of interest.

RECOMMENDATIONS

Develop a nationwide network of advanced biomanufacturing hubs to support innovation and commercialization for advanced biotechnologies. Each hub would provide manufacturing capacity to biotech firms making the transition from proof of concept to producing early commercial amounts. The government would procure or partially cover the cost of biomanufacturing equipment (such as utilities, labs, reagents, and storage), ease regulations for private companies building out larger facilities on the same campus, and colocate federal research programs where practical. The private sector would commit to investing a multiple of federal funding into each hub to support related infrastructure and activities, such as raw material processing, quality control, and workforce training. That network could begin with four to six hubs and scale further as performance, resources, and national security conditions warrant.

Establish a biotechnology investment fund to support high-priority areas of national security technology that are left unaddressed by current initiatives, along the lines proposed by the National Security Commission on Emerging Biotechnology.[76] The fund would be professionally managed by a nongovernmental investment partner and could provide equity financing to early-stage biotech start-ups to cover a portion of preclinical costs and target drug developments in areas such as mRNA platforms, genome engineering, and synthetic biology. The fund could also leverage capital from allies, such as the EU, Japan, and South Korea, and gains from investments could be reinvested, thereby reducing the need for future public funding.

Offer advance market commitments to clearly signal demand for biotech applications important to national security. As the success of Operation Warp Speed highlights, guaranteeing government purchases encourages biotechnology and pharmaceutical companies to invest and innovate. Historically, these demand-side programs have focused on addressing public health emergencies, but their scope should be expanded to include national security applications, such as treatments for rare diseases with high potential to be weaponized and systems that can provide rapid response to infectious and viral diseases. The U.S. government would have to state clearly what it wants the private sector to develop, along with sourcing requirements, and commit to buying if certain safety and efficacy thresholds are met.

Shift contract development and manufacturing organizations and contract research organizations services to trusted markets. The United States should explore deepening partnerships that would encourage CDMO/CRO services to shift from China to more trusted markets. Canada, the Czech Republic, India, Poland, South Korea, and the UK all have capabilities that could enhance U.S. resilience. The United States can work with partners to diversify sourcing, establish joint manufacturing facilities, and create shared regulatory standards for quality and traceability, strengthening shared biotech ecosystems while mitigating overreliance on China. U.S. and like-minded partners could rapidly mobilize jointly owned capabilities abroad in the event of a biothreat or lack of access to critical technology. Those efforts could also increase U.S. government visibility into concentration risks.

Partner with U.S. companies to stockpile KSMs and APIs for essential medicines and support production in trusted markets. Stockpiling is feasible because APIs have a shelf life of one to five years, and KSMs can remain viable for even longer. The Phlow Corporation, a public benefit corporation that is mandated to balance profit and social impact, has received government funding over ten years to establish the first API stockpile, two advanced manufacturing facilities, and a national storage network. Building on these efforts, Congress should dedicate additional funding for the executive branch to incentivize U.S. pharma companies to build a six-month emergency supply of KSMs and APIs for essential medicines and mandate that stockpiled materials come from trusted sources. Most large U.S. pharma companies have storage facilities that could be expanded, so the marginal costs are not high, and the U.S. government could cover the incremental overhead, storage, and working capital needed for companies to hold reserve supply. The national stockpile could be retained and focused to fill any remaining gaps.

Add a reporting requirement for outbound investments to countries of concern in biotechnology. The investments tracked could include, for example, U.S. and multinational biopharmaceutical partnerships, licensing deals, co-development arrangements, and research collaborations that transfer critical IP or support capability building in areas such as genomics, novel biology, and AI or computational biology. That information would help the U.S. government better assess current risks and consider whether controls are needed. It could also reveal opportunities to work with partners and allies, whether in sharing information and limiting investment from undesired destinations or increasing cooperation on R&D.

Critical Minerals

The need for critical minerals cuts across AI, quantum, and, to a lesser extent, biotech. The Trump administration has taken several major steps toward developing domestic sources of critical minerals. On President Donald Trump's first day back in office, he declared a national energy emergency and identified critical mineral shortages as a direct threat to U.S. economic and national security. Another executive order, Unleashing American Energy, aims to cut red tape and accelerate the permitting, leasing, and development of domestic energy and mineral resources. More recently, the Department of Defense announced an innovative partnership with MP Materials that uses a range of public-private tools, including the U.S. government taking an equity stake and providing a price floor commitment, loans, and offtake guarantees, alongside private financing.

The recommendations below are intended to build on this progress with the ambitious goal of sourcing no more than 65 percent of each critical mineral, at any stage of processing, from a single country by 2035 (see figure 11). Many of these recommendations, particularly those related to stockpiling and mining, will require investing in supporting infrastructure to facilitate extraction, transportation, and storage. Deepening cooperation with partners and allies that have significant resources and capabilities can advance these recommendations. For example, the Critical Minerals Action Plan from the Group of Seven (G7) calls for working together and with partners beyond the G7 to build standards-based markets, mobilize capital, and promote innovation.[77] The EU, Japan, and Norway are promising partners for collaboration in recycling, deep-sea mineral collection capabilities, and material substitution. Others, such as Australia and Canada, have important capabilities for accelerating the development and deployment of advanced recovery methods, processing, and refining. Additionally, joint efforts in emerging markets could support the exploration, processing, and stockpiling of key materials.

INCREASE SUPPLY AND PRODUCTION

Expand the National Defense Stockpile. Congress should provide $2 billion to scale the National Defense Stockpile (NDS) to secure high-risk materials essential to AI, quantum, and biotech infrastructure. While some materials are already held in limited quantities, others—such as gallium, neodymium, holmium-copper 2, helium-3, niobium, and grain-oriented electric steel—are not stocked or exist only in trace amounts. The NDS is designed to ensure access to key inputs during national emergencies, but it currently lacks the

FIGURE 11

Where Do Critical Minerals for AI and Quantum Come From?

U.S. import share for critical minerals where the United States is at least 65% dependent on a single country

Leading supplier: ■ China ■ Brazil ■ Mexico

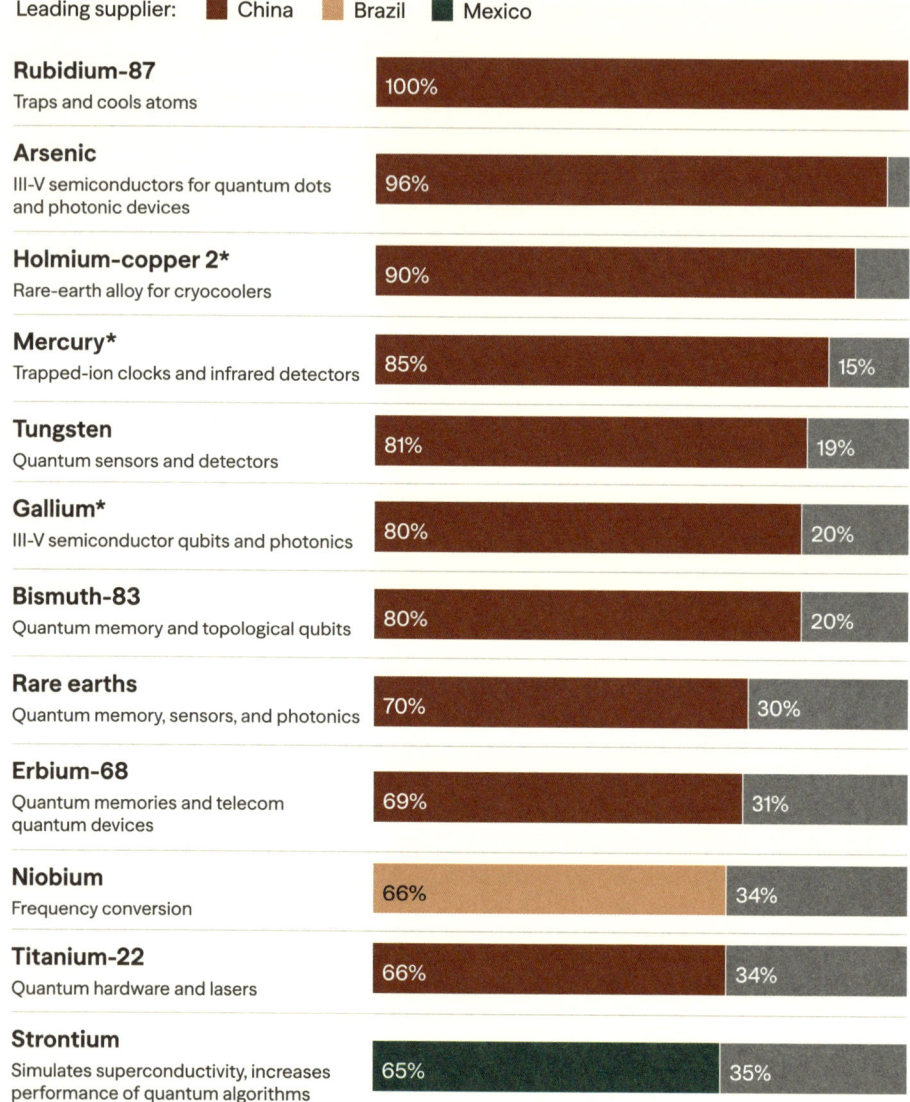

Rubidium-87
Traps and cools atoms — 100%

Arsenic
III-V semiconductors for quantum dots and photonic devices — 96%

Holmium-copper 2*
Rare-earth alloy for cryocoolers — 90%

Mercury*
Trapped-ion clocks and infrared detectors — 85% / 15%

Tungsten
Quantum sensors and detectors — 81% / 19%

Gallium*
III-V semiconductor qubits and photonics — 80% / 20%

Bismuth-83
Quantum memory and topological qubits — 80% / 20%

Rare earths
Quantum memory, sensors, and photonics — 70% / 30%

Erbium-68
Quantum memories and telecom quantum devices — 69% / 31%

Niobium
Frequency conversion — 66% / 34%

Titanium-22
Quantum hardware and lasers — 66% / 34%

Strontium
Simulates superconductivity, increases performance of quantum algorithms — 65% / 35%

*Values for mercury, holmium-copper 2, and gallium are approximated due to lack of data on global production and U.S. imports.

Source: U.S. Geological Survey

resources and flexibility to meet projected demand across those strategic sectors. Congress should provide new appropriations and modernize stockpile authorities, such as annual purchase caps, to enable timely acquisition of critical materials. It should also explore more creative means of acquiring materials, as China has been using licenses to ensure buyers cannot increase rare earths purchases to stockpile.

Map untapped resources. Congress should provide $1.6 billion to complete the U.S. Geological Survey Earth Mapping Resources Initiative to identify and assess domestic critical mineral reserves across the continental United States. High-resolution geological surveys—augmented with satellite imagery and AI-driven subsurface tomography—could create a "Google Earth" for underground resources. Full mapping would take over fifteen years, while mapping only priority areas would take approximately six years and require up to $650 million.

Accelerate permitting. Reform federal permitting to shorten the process of bringing mines online, as the United States currently has the second-longest lead time—an average of twenty-nine years. Recent efforts to accelerate permitting include President Trump's Executive Order 14241, which in March 2025 added critical minerals infrastructure projects to the FAST-41 program, a process for improving federal agency coordination and timeliness of environmental reviews for infrastructure projects. Additional complementary reforms could include working with Congress to consolidate overlapping permitting requirements across regulatory bodies, increasing resources for permitting agencies (Bureau of Land Management, Forest Service), and clarifying provisions of the General Mining Law to reduce litigation-related delays.

INNOVATE BEYOND MINING

Establish robust recycling infrastructure for end-of-life batteries, electronics, and permanent magnets, especially high-value materials like lithium, cobalt, and rare earths, by increasing financial incentives for R&D and recycling and supporting collection logistics through public-private partnerships. The DOE's ReCell program has made progress by partnering with industry to develop cost-competitive methods for recycling advanced batteries.

Pursue advanced recovery methods by investing in next-generation separation techniques, such as synthetic biology and bioleaching, which use microorganisms to extract valuable metals from waste. Funding should be increased for initiatives with proven success, like the DOE's Fossil Energy and Carbon Management program, the MINER program of the Advanced Research

Projects Agency-Energy (ARPA-E), and DARPA's EMBER program. Additional funding to universities and NSF programs conducting research into emerging extraction techniques could also speed up development of such methods.

Invest in substitution technologies by supporting research into engineered alternatives, such as metamaterials and quantum alloys that could reduce or eliminate the need for rare earths in key applications. ARPA-E's REACT program is developing cost-effective alternatives to rare earths and could be further scaled and expanded in scope to include alternatives beyond magnet materials. Once these alternatives are identified, additional investment will be needed to create the manufacturing capacity to produce meaningful volumes.

Workforce

The Trump administration's America's Talent Strategy recognizes that a skilled workforce is essential to leading in key technologies and achieving U.S. economic security and calls for action across five pillars: industry-driven strategies, worker mobility, integrated systems, accountability, and flexibility and innovation.[78] As these efforts are operationalized, a natural next step is to strengthen education and training pathways for AI, quantum, and biotechnology.

When people think about AI, biotech, and quantum, they might picture PhD-trained coders sitting at their keyboards or scientists wearing lab coats. Without question, the United States needs to ensure it is training the world's top STEM talent. Equally essential are the technical trades workers who can build, run, and maintain the physical infrastructure that makes these industries possible. Building and running AI data centers, for example, will require training more electricians to manage power needs, machinists to fabricate key components and equipment, and technicians to install and maintain cooling systems.

Solutions will require collaboration across the U.S. innovation ecosystem: federal, state, and local government, early to graduate education, vocational and research programs, and private businesses of all sizes. Progress is already underway at the state and local levels, in places such as Montana's Photonics and Quantum Alliance, North Carolina's Biotechnology Center Landing Pad, the University of Florida's Sid Martin Biotech incubator, and Illinois's Quantum and Microelectronics Park. While each example is unique, all are harnessing partnerships between universities, the private sector, and government.

Given the importance of this topic, which extends far beyond the focus of this report, the Task Force recommends that it be the subject of a dedicated study that builds upon the administration's America's Talent Strategy and recent work. In 2018, CFR's Task Force on workforce development, *The Work Ahead: Machines, Skills, and U.S. Leadership in the Twenty-First Century*, made a range of recommendations for the federal government, state and local governments, and the private sector to rebuild the links among work, opportunity, and prosperity for all Americans. Many of these recommendations remain relevant today, while others should be updated in light of rapid technological change and intensified geopolitical competition.[79]

COMPETING IN THE AGE OF ECONOMIC WARFARE

The age of economic warfare has arrived, yet the United States is still wielding last century's weapons. Many U.S. economic tools were designed during a period of intense geopolitical competition, but innovation and China's rise pose new challenges that extend beyond the emerging technologies examined in previous sections. The U.S. government needs better coordination across agencies, deeper technical expertise, and a more proactive approach to partnering with the private sector to successfully shape and execute economic security policies. Strikingly, unlike in the realm of military power, there is no "doctrine" for using economic security tools, no guiding principles, and few coordinated drills and trainings. To fill these gaps, the Task Force recommends establishing an Economic Security Center, upgrading economic security tools, and adopting principles to strengthen the practice of economic security.

Establish an Economic Security Center

The U.S. government needs a central hub for economic security. Capabilities for economic security are spread across several departments and agencies, and consequently, U.S. economic security policy has suffered from fragmented decision-making and a lack of institutional capacity. Effective economic security policy also depends on clear, sustained communication between government and the private sector, which historically has been too ad hoc. To address these challenges, the executive branch and Congress should work together to create an Economic Security Center at the Department of Commerce to serve three roles:

A U.S. government coordinator that leads

- **economic war games** that bring together company representatives and government officials to improve the use of defensive and offensive tools, including by simulating responses to external shocks, and to explore ways to take advantage of opportunities in foreign markets;
- **an interagency committee on industrial policy** that identifies strategic industries and works to remove barriers and align incentives across the U.S. government to support industrial goals, playing a role for industrial policy similar to the coordinating role that the Treasury Department plays for CFIUS;

- **an export control task force** with the Department of Treasury and Department of Justice that works to increase the effectiveness of export control penalties as a deterrent, harnessing modern data analytics to monitor controls and identify violations and mobilizing the full range of U.S. government authorities to impose penalties; and
- **enhanced impact assessments of foreign investment,** including through collaboration with the Treasury Department to integrate a cost-benefit analysis of CFIUS and outbound actions into CFIUS's annual report that assesses the cost of compliance and impact of rules on R&D, investment, and talent in tech.

A technical advisor who provides

- **on-demand analysis and advice** on critical and emerging technologies from experts, recruited through flexible hiring authorities, with whom designated offices across the U.S. government can consult;
- **a fast technology teardown program,** managed in partnership with the Department of Defense and the National Security Agency;
- **a supply chain forecasting and monitoring system** that collects information from companies on supply chain risks, including through the Defense Production Act, and identifies emerging risks and dependencies of concern before they become acute; and
- **regular assessments of barriers to innovation,** such as legal and regulatory processes and prohibitions, and recommendations for alleviating these burdens that are shared with the interagency committee on industrial policy.

A private-sector partner that offers

- **expanded protections** for U.S. companies sharing information with the government on economic security issues, potentially through legislation similar to the Cybersecurity Information Sharing Act of 2015, which is currently being updated and has successfully encouraged the private sector to share cyber threat information;
- **data** to assist U.S. companies with due diligence of foreign entities, including investors, customers, and research partners for priority technologies;
- **two-way communication channels,** allowing the government to share information on risks and opportunities with relevant groups of companies and the companies to share information securely with the government in real time;

- **exemptions for designated critical technology projects from National Environmental Policy Act review,** which could require approvals similar to those in the Fiscal Responsibility Act of 2023;[80]

- **a guide to relevant U.S. government programs,** including services offered, requirements, examples of success cases, and points of contact to help small and medium-sized businesses access U.S. government programs; and

- **stronger U.S. government advocacy for outbound investment into certain sectors.** At present, U.S. government advocacy prioritizes exports and inbound investment, leaving U.S. investors in strategically important foreign projects at a disadvantage.

Upgrade Economic Security Tools

U.S. economic security tools need to be modernized. Export controls and industrial policy require the most immediate attention, given their importance to ensuring U.S. technological leadership and the challenges identified in previous sections. To be sure, other tools require improvement as well but have been the focus of more recent reforms.[81] Upgrading export controls and industrial policy as recommended below would enhance the defensive and offensive aspects of U.S. economic security, respectively.

On the defensive side of the U.S. economic tool kit, expansion of export controls has outpaced the ability of the U.S. government to effectively enforce them, as highlighted by U.S. semiconductor controls. Complying with an increasingly complex control regime can disadvantage U.S. companies, especially smaller firms, and delay the delivery of important capabilities to partners and allies. Overall, the U.S. government should move to a more selective, powerful posture of fewer controls with higher penalties for violations.

To make export controls more effective, the Task Force recommends the following actions:

- **Conduct a regular review of restrictive measures, including export controls, to determine if the control should be retained or retired.** That assessment, which could be completed every four years or so, should have a public comment period. Narrowing the scope to fewer, more critical items would increase the focus of enforcement authorities and reduce costs to U.S. firms competing abroad. Dual-use items that are widely available globally should be removed from U.S. controls. Global availability can result naturally when technologies mature, or if technology controls are not effectively multilateralized.

- **Establish a central foreign military sales–only list that is reviewed and updated every two years.** The current process is too fragmented across services and agencies, too broad in scope, and too slow to remove items. A single foreign military sales–only list should focus on technologies that require training and services for long-term interoperability and sustainment, encouraging direct commercial sales where possible to improve contracting speed. The State Department would maintain the list, in coordination with the DOD, and could share it privately with cleared U.S. industry representatives. Regular updates would allow the removal of items from the list to better reflect technological changes.

- **Exempt U.S. allies from select license requirements by applying lessons from AUKUS.** While even AUKUS, the trilateral security agreement among Australia, the United Kingdom, and the United States, could be further enhanced, it has removed a significant number of export permit requirements among the three countries. The United States should offer to remove license requirements in exchange for allies more closely aligning their export control regimes. Making progress with additional allies would have the added benefit of allowing the U.S. government to devote more time toward managing controls on countries of concern.

- **Transfer approval for U.S. government advocacy on defense sales from the Commerce Department to the State Department**, which already oversees international security cooperation and could move faster on requests to support partner needs. That adjustment would have the added benefit of allowing the Commerce Department to focus on supporting nondefense activities, including giving additional attention to outbound investment. The Defense Department would continue to evaluate partner requirements, oversee contracts, and provide training, logistics, and sustainment to ensure U.S. systems are transferred and integrated effectively.

- **Pilot defense "deal teams" in partner markets.** This would bring together key staff at U.S. embassies from the Departments of Defense, Commerce, and State on a more regular basis to identify and advance priority foreign military sales.

On the offensive side of the U.S. economic tool kit, upgrading several aspects of U.S. industrial policy would help mobilize larger pools of private capital. The United States benefits from the world's deepest capital markets, but as highlighted in the quantum and biotechnology sections, private investment flows toward areas with the highest perceived returns. In strategically important areas where risks are higher, or potential returns are lower, the U.S. government has an important role to play in mobilizing private capital.

To catalyze capital toward underinvested areas, the Task Force recommends the following actions:

- **Create a "fast track" for investment from trusted foreign sources.** That could include updating and expanding the list of "excepted" states and investors (beyond the Five Eyes partners) and replacing the board member nationality test with an assessment of ties to U.S. adversaries. Trusted investors would avoid mandatory filings and benefit from streamlined filings. CFIUS would retain its ability to review concerning transactions.

- **Reauthorize the U.S. International Development Finance Corporation with several enhancements.** Those include updating its budgetary treatment of equity to mirror existing federal lending programs, aligning country eligibility with World Bank lending categories and expanding country eligibility to include middle- and even certain high-income countries, and extending the duration and size of political risk insurance products. Those updates would allow the DFC to do more with its own capital and to catalyze more private capital, in more strategically important markets.[82]

- **Reauthorize the U.S. Export-Import Bank with several enhancements.** Those include raising its default rate cap from 2 to 4 percent, exempting national security transactions from the default cap, replacing domestic content requirements with requirements based on export value and job creation, expanding the China and Transformational Exports Program to cover other important sectors, such as data centers and critical minerals, and authorizing financing for defense exports to foreign military sales–eligible countries by initially making available up to 10 percent of its portfolio for defense exports with the ability to increase this cap to reach 20 percent over five years, if demand warrants.

- **Offer loan guarantees** for critical technology start-ups through a federally backed program modeled after Veterans Affairs loans and issued by the Defense Department's Office of Strategic Capital. Eligibility criteria could include designated critical technology areas (semiconductors, quantum, biotech) and minimum thresholds for technical readiness. Those loans would be in the form of non-dilutive capital, which does not require founders to give up equity or ownership.

- **Preserve qualified small business eligibility** for hard technology start-ups by increasing the gross asset threshold from the existing $75 million to $125 million, excluding specific asset classes (federally funded R&D assets or non-dilutive grants) from counting toward the limit, or deferring the test to a later stage, such as revenue or technical milestones.

Guide Economic Security Actions With Principles

The economic security needs of the United States are continuously evolving, and much remains to be done to establish trust in supply chains. Looking beyond today's immediate challenges, the following principles are intended to help U.S. policymakers decide when and how to intervene in markets in the name of national security. Answering the questions below in the affirmative should increase one's confidence in the success of an economic intervention for national security purposes. Answering in the negative should give one pause and encourage additional consideration. To further assist decision-making, each principle also includes advice from former senior U.S. officials.

Market impediments: *Is there a market impediment that could harm U.S. national security?* Market impediments that affect national security include the following:

- **Dual-use technologies,** capital, or other forms of support that can benefit an adversary's military or provide an adversary with advantages during a conflict.

- **A monopoly or geographic overconcentration** of critical supply chains, especially but not limited to countries of concern, whose governments could eventually weaponize such chokepoints against the United States and its allies.

- **A public goods problem,** such as a lack of private investment in areas that are strategically important (e.g., defense, critical infrastructure, and advanced high-value manufacturing) but commercially risky or uneconomical.

- **Asymmetric information,** such as firms or foreign governments being unaware of an entity's connections to an adversary.

Advice from practitioners:

- Differentiate among sources of supply (trusted vs. non-trusted) and among goods (strategic vs. nonstrategic) to keep the focus on national security.

- Clearly and publicly explain the threat to national security, which will help marshal domestic and international support for successful implementation.
- Share relevant intelligence with allies, which increases trust, builds confidence in the national security justification, and dispels suspicions of ulterior motives.

Focus: *Are the objectives for the intervention clearly defined and clearly related to national security, and do the tools directly target the risk?* Past practice suggests that U.S. economic tools, used appropriately under certain conditions, have successfully advanced a range of objectives, including

- **degrading** an adversary's military capabilities;
- **delaying** an adversary from acquiring dual-use technologies;
- **de-risking** supply chains of critical goods;
- **deepening** military-to-military ties and interoperability with allies;
- **defending** against adversarial control or influence over sensitive assets; and
- **deterring** or punishing aggression and other violations of international norms.

Advice from practitioners:

- Articulate the behavior being penalized and the outcomes expected when economic pressure is combined with diplomacy, military force, or other forms of power.[83]
- Limit the use of the most potent tools—including broad-based financial sanctions and export controls—to situations in which vital U.S. national security interests are at stake.
- Consult with partners and allies to refine objectives that advance common security interests, when possible, and begin building international support from the start.
- Avoid purely symbolic uses of economic security tools, which can signal a lack of commitment to an adversary and undermine U.S. economic power.
- Take steps to mitigate negative economic spillovers in advance. For example, if potent sanctions require risking a temporary disruption to energy markets, devise a plan to stabilize the markets.

Feasibility: *Are key conditions favorable for the U.S. government intervention to succeed?* Important factors include the following:

- **Capability**, such as control over a key network, input, or know-how; sufficient resources for investment; and clear legal authority.
- **Non-substitutability** to limit the extent to which targets can successfully adapt to U.S. interventions over the long term.
- **Bipartisan support** for ensuring the durability of economic security interventions, which can take many years to achieve their intended effects.
- **Government capacity** for implementation, including technical expertise, data and intelligence, technology, enforcement capabilities, and authorities to navigate regulatory obstacles.
- **International support** in certain circumstances to maximize the intervention's effectiveness, prevent circumvention, and preserve the foundations of American economic power.
- **Time** for implementation, accounting for key actors' actions and reactions, political and economic conditions in adversarial nations, and trends such as technology trajectories.

Advice from practitioners:

- Identify the biggest obstacles to implementation so additional capabilities and authorities can be requested, if needed, at the beginning of the process.
- Ensure the threat of imposing and sustaining punitive measures is credible despite inevitable costs.
- Consult regularly with the private sector, especially for insights on technology and supply chains, understanding that companies have their own interests.
- Assume adversaries will adapt and consider whether the amount of time it will take the adversary to overcome the intervention is meaningful.
- Prepare for escalation so that measures can be ratcheted up as the adversary adapts. Static sanctions and export controls wane in impact over time, as the adversary adjusts while restrictions remain stable. The only way to maintain pressure is to proactively tighten restrictions.

Trade-offs: *Do the benefits of the intervention outweigh the costs, and is the intervention proportionate to the threat?* Relevant costs include the following:

- **Risk of retaliation** from adversaries, including both economic and non-economic measures.
- **Economic costs** to U.S. households and companies and broader market impacts, depending on the intervention's scale and scope.
- **Innovation costs,** including the ability of U.S. companies to invest in R&D and effects on U.S. leadership and competitiveness in science, technology, engineering, manufacturing, and other areas.
- **Diplomatic costs,** including how economic impacts on partners and allies could affect their support over time for the intervention and related issues.
- **Systemic costs,** including weakening U.S. capabilities for economic leverage during future contingencies.

Advice from practitioners:

- Adopt a more selective and powerful posture by raising the threshold for using primary sanctions and lowering the threshold for using secondary sanctions. That would narrow sanctions usage to the most important cases while increasing their impact and reducing incentives for the use of non-U.S. alternatives.
- Err on the side of overestimating what is required to achieve the intended policy effects, given the gradual nature of economic power and the ability of adversaries to adapt.
- Reevaluate costs and benefits as new facts emerge about the effectiveness of the measures, as well as their consequences for U.S. competitiveness.

Those principles are a starting point for moving toward a more systematic approach to the practice of economic security. Using them will help ensure that U.S. government interventions are more likely to achieve their objectives. It will remain essential for the U.S. government to assess the impact of its interventions, both to adjust when needed to achieve the objective at hand and to inform the improvement of U.S. economic security institutions, tools, and practices.

The race for tomorrow's foundational technologies is underway, and China is gaining ground. The difference between leading and following in AI, biotech, and quantum will be measured in trillions of dollars in economic opportunity, as well as control over critical supply chains and defense capabilities. The United States must marshal its economic power to stay ahead.

APPENDIX

This section provides additional information on this report's top recommendations, as well as granular supply chain dependencies for each tech area.

TABLE I

Supply Chain Risks for AI

SUPPLY CHAIN SUBCATEGORY	SELECTED COMPONENT/ PROCESS (NIR = NET IMPORT RELIANCE)	DESCRIPTION	PRIMARY SOURCES OF SUPPLY	ALTERNATIVE SOURCES
Data Center Components	Diesel backup generators and uninterruptible power supply (UPS)	Alternative power source	UK (24%)	BR, AE, DE
Data Center Components	Other power equipment (converters, inductors, high- and low-volume transformers, data processing units)	Ensures smooth transmission and distribution of power from grid to data centers	CN ($3.3B), MX ($776M)	TH, JP, KR CA
Data Center Components	Battery energy storage systems	Stores energy for later use	Minerals (CD, ID, CN); batteries and infra (CN)	AU, AR, CL, CA, RU
Data Center Components	Printed circuit boards	Supports data center operations	CN (30.3%)	TW, JP, KR
Data Center Components	Processing and data storage capacity	AI-oriented data centers require much higher data storage capacity	Current global AI data center IT demand: 44 GW/yr; 2030 projection: 156 GW/yr	
Data Center Components	Next-gen liquid cooling technologies	Manages heat and energy consumption	TW ($13B), MX ($711M), DE ($216M)	CA, CN
Data Center Components	Networking equipment (transceivers, laser diodes modems, switches, routers, servers, CPUs)	Facilitates data flow, security, and connectivity, both intra-DC and to end products	Top 11 suppliers have 80% share; 6/11 are CN. Tail suppliers (56% CN) lead in legacy	TW, TH, NL
Semiconductors	Logic chips (<7nm)	Advanced microchips for processing and execution	TW (>70% of global capacity)	KR
Semiconductors	Foundry services	Fabrication capacity bought by "fabless" firms	TW (60% of global share)	KR, CN
Semiconductors	Back-end assembly, testing, and packaging sites and supply chain	ATP capacity purchased by "fabless" firms to reduce lead time and cost	TW (52% of global capacity); 85% of sites in Asia: CN (33%), TW (6%)	CN

TABLE I (CONTINUED)

SUPPLY CHAIN SUBCATEGORY	SELECTED COMPONENT/ PROCESS (NIR = NET IMPORT RELIANCE)	DESCRIPTION	PRIMARY SOURCES OF SUPPLY	ALTERNATIVE SOURCES
Semiconductors	Chemicals and gases (ultrapure wet chemicals, dry etchants, photoresists—HF, WF6, Ne, CeCO3F, CePO4)	Critical for wafer manufacturing, cleaning, and chip etching process	CN (60%–80%)	KR, TW, JP
Semiconductors	IC substrates	Connect wafer to PCB	TW (33%)	KR (25%), JP (19%)
Semiconductors	Semiconductor manufacturing equipment	Chip manufacturing process (etching, lithography, deposition)	JP, NL (90%+)	SG, MY
Critical Minerals	Arsenic (100% NIR)	Dopant for silicon chips	CN (96%)	JP (3%)
Critical Minerals	Cobalt (76% NIR)	Enhances performance of electronic components	NO (27%)	FI (17%), JP (14%)
Critical Minerals	Indium (100% NIR)	Used in optical communications	KR (29%)	JP (18%), CA (14%)
Critical Minerals	Gallium (100% NIR)	Key for compound chips, outperforms silicon	JP (24%)	CN (19%), DE (19%)
Critical Minerals	Germanium (>50% NIR)	Enhances performance of high-speed chips	BE (42%)	CN (23%), CA (23%)
Critical Minerals	Palladium (36% NIR)	Used against corrosion and in ceramic capacitors	RU (32%)	ZA (32%), BE (8%)
Critical Minerals	Silicon (<50% NIR)	Base material for chips	BR (38%)	CA (28%), NO (13%)
Critical Minerals	Rare earths (neodymium, scandium, yttrium) (80% NIR)	Polishing wafers, lasers; insulation for advanced chips	CN (70% rare earths, 99% heavy rare earths)	MY (13%), JP (6%)
Critical Minerals	Tungsten (>50% NIR)	Aids transistors' heat resistance and durability	CN (27%)	DE (14%), BO (8%)

Note: Areas beyond the data center, such as upgrading supporting U.S. power and electricity infrastructure, including through developing new technologies such as fusion, are important but beyond the scope of this case study.

Source: CFR research

TABLE II

Supply Chain Risks for Quantum Technologies

SUPPLY CHAIN SUBCATEGORY	SELECTED COMPONENT/ PROCESS	DESCRIPTION	PRIMARY SOURCES OF SUPPLY	ALTERNATIVE SOURCES	MODALITY
Control	Control electronics (cryogenic CMOS for photons and topological)	Manages quantum operations and manipulates qubits	IL (quantum machines), AU (Q-CTRL)		All
Control	Precision lasers	Controls quantum qubits, laser-cools atoms	CN (Precilaser)	US (Vescent, coherent)	Trapped ion, neutral atom, photonic
Control	Blue gallium nitride laser diodes	Cools and traps atoms	JP		Neutral atom
Components	Superconducting coaxial cables	Enables ultra-low-loss transmission of signals		JP, UK, CN	Superconducting, spin qubits
Components	Superconducting nanowire single-photon detectors	Detects photons for communication and computing	DE, IT	FR, SE	Trapped ion, photonic
Components	Semiconductor saturable absorber mirror	Key component for optical frequency combs/lasers	FI	N/A	Neutral atom, trapped ion, photonic
Components	Laser diodes	Traps and cools atoms	DE (TOPTICA)		Neutral atom, trapped ion, photonic
Components	High electron mobility transistor amplifiers	Allows for fast signal processing and low noise	SE		Superconducting
Components	200 mm sapphire wafers	Substrate for fabricating qubits	RU, JP		Superconducting, trapped ion
Environment	Cryocoolers and compressors	Provides initial stages of cooling	CN, US (Cryomech), FN	DE (Kiutra), CA (Zero Point Cryogenics), JP	Superconducting, trapped ion
Environment	Dilution refrigerator	Creates ultra-low temperatures	FN	UK, NL, US (Maybell)	Superconducting, quantum dots, photonic, topological qubits
Environment	Ultra-high vacuum chambers and components	Creates clean and controlled environments	JP (ULVAC), DE, UK for components		Trapped ion, neutral atom
Minerals	Helium-3	Used in dilution refrigerators	US (Interlune), RU	Moon mining, CN	Superconducting, photonic, topological qubits
Minerals	Holmium-copper 2	Rare earth alloy critical for cryocoolers	CN (90%) (AEM rare earths)	JP	Superconducting
Minerals	Niobium	Key for frequency conversion	BR, CN	N/A	Photonics, communication
Minerals	Silicon-28	Substrate for spin qubits, quantum dots	BR (38%)	CA, NO, AU	Spin qubits, quantum dots, photons

Appendix

TABLE II (CONTINUED)

SUPPLY CHAIN SUBCATEGORY	SELECTED COMPONENT/ PROCESS	DESCRIPTION	PRIMARY SOURCES OF SUPPLY	ALTERNATIVE SOURCES	MODALITY
Minerals	Rubidium-87	Traps and cools atoms	CN (>90%)	DE, RU (minimal)	Neutral atom
Minerals	Aluminum	Superconducting qubits	CA (56%)	AE, BH, CN	Superconducting
Minerals	Rare earths (80% NIR)	Used in quantum memory, sensors, and photonics	CN (70%)	MY, JP	Photonic
Minerals	Strontium (100% NIR)	Act as qubits and increase fidelity of operations	MX, DE	IR, ES, CN	Neutral atom
Minerals	Tantalum (100% NIR)	Acts as thin-film superconductor on silicon chip	CN (22%)	AU, DE, ID, JP, KZ	Superconducting / cat qubit
Minerals	Indium (100% NIR)	Used in photonic semiconductors and qubit chips	KR (29%)	JP, CN, CA, BE	Superconducting, photonic
Minerals	Bismuth-83	Used in quantum memory and topological qubits	CN (~80%)	LA (~10%), SK (~5%), JP (~2%)	Spin qubits, topological qubits
Minerals	Arsenic-33	Integral to III-V semiconductors for quantum dots and photonic devices	PE (~45%), CN (~40%), MA (~13%)	BE, RU (minor)	Photonic, semiconductor spin qubits
Minerals	Barium-56	Used in trapped-ion qubits and nonlinear optical crystals	IN (~30%), CN (~22%), MA (~15%)	KZ (~6%), US (~5%), TR (~3%)	Trapped ion, photonic
Minerals	Titanate – (Ti = 22) (note: titanate is a compound of Ti and O)	Ferroelectric perovskites for quantum photonics and transducers	Synthesized from Ti and Ba/Sr compounds	–	Photonic, superconducting
Minerals	Antimony – 51	Used in compound semiconductors for infrared quantum sensors and lasers	CN (~48%), TJ (~25%)	TR (~7%), RU (~5%), MM (~6%)	Topological qubits, photonic and IR sensors
Minerals	Erbium – 68	Used in quantum memories and telecom quantum devices	CN (~69%), MM (~11%), US (~12%)	AU, TH, VN	Photonic, spin qubit
Minerals	Gallium – 31	Critical for III-V semiconductor qubits and photonics	CN (~80%+)	JP, RU, KR (each <5%)	Photonic, spin qubit, quantum sensors
Minerals	Germanium – 32	Used in semiconductor quantum technologies and photodetectors	CN (~60%); CA, FI, RU, US (~40% combined)	BE, DE (minor)	Photonic, spin qubits, quantum sensors
Minerals	Carbon – 6	High-purity diamond used for quantum spin qubits and sensors	RU (~35%), BW (~21%)	CA (~13%), CD (~8%)	Spin qubits and sensors, photonic

TABLE II (CONTINUED)

SUPPLY CHAIN SUBCATEGORY	SELECTED COMPONENT/ PROCESS	DESCRIPTION	PRIMARY SOURCES OF SUPPLY	ALTERNATIVE SOURCES	MODALITY
Minerals	Mercury – 80	Utilized in trapped-ion clocks and infrared detectors	CN (~80 - 90%)	TJ (~5%), MX (~2%)	Trapped ion, photonic/detection
Minerals	Molybdenum – 42	Used in superconducting quantum devices and detectors	CN (~40%), CL (~18%)	PE (~15%), US (~14%)	Superconducting, photonic, topological
Minerals	Titanium – 22	Important in quantum hardware and lasers	CN (~66%)	JP (~17%), RU (~7%)	Superconducting, photonic
Minerals	Tungsten – 74	Employed in quantum sensors and detectors	CN (~81%)	VN (~4%), RU (~3%)	Photonic/detector, superconducting

Source: CFR research

TABLE III

Supply Chain Risks for Biotechnology

SUPPLY CHAIN SUBCATEGORY	SELECTED COMPONENT/ PROCESS	DESCRIPTION	PRIMARY SOURCES OF SUPPLY/ BOTTLENECK	ALTERNATIVE SOURCES
Drug Inputs	Generic active pharmaceutical ingredient	Raw material input that delivers the intended therapeutic effect (e.g., acetaminophen)	IN (>50% of U.S. supply), CN (1/3 global capacity), only 10% made in United States by three firms	MX, EU
Drug Inputs	Key starting materials	Structural building block for APIs to synthesize	India depends on Chinese KSMs (>50%), ~80% of U.S. drug KSMs sourced from China	Domestic: U.S. subsidies proposed in BARDA, national biotechnology and biomanufacturing initiative; international: MX, Eastern Europe
Biomanufacturing	R&D outsourcing (CROs and CDMOs)	Contracted research and development and manufacturing organizations	~80% of global CDMO capacity in Asia and Europe; 79% of companies tied to Chinese CDMOs	IN, Eastern Europe, some U.S. CDMO incubators proposed
Biomanufacturing	Fermentation/ bioproduction	Cultivation of biologics via microbial or cell culture in bioreactors	Medium-sized facilities $2B+, six years to build; long lead time for complex bioreactors	Recent developments in single-use bioreactors with shorter lead times and lower investment
Biomanufacturing	Genetic manufacturing	Tools and facilities for gene/cell therapy	12- to 15-month wait time for GMP-grade viral vectors; GMP facility cost; limited CMO bandwidth	United States lags China (40% of global cell therapy trials); EU slow due to regulation
Innovation	Speed to market	Time from discovery to product approval	U.S. avg is 8 yrs; China leads with 23% of global innovation; 1/3 of global licensing from Chinese biotechs in 2024	United States leads in R&D spending, China leads in trial volume; FDA approval time needs lowering
Regulatory/IP	Data security/adversary access risk	U.S. biotechs dealing with sensitive data and information (e.g., human genomes)	Lack of laws and cybersecurity measures protecting against risk	United States devotes more money and builds legislation around data security for bioinformation

Source: CFR research

ADDITIONAL VIEWS

Although the report lays out the stakes and vulnerabilities related to economic security, the importance of federal funding for basic research in technological leadership should be emphasized.

A pipeline of technological breakthroughs in the United States has been federal funding for basic research, including through agencies such as the National Science Foundation, the National Institutes of Health, the Centers for Disease Control and Prevention, the Environmental Protection Agency, DARPA, and the Departments of Energy and Agriculture. Because of the unpredictability of its results, its long time horizons, and the open-source sharing of its ideas, most basic research is funded by the public sector around the world. In the United States, the federal government has long been the largest source of funding for basic research, and most of this research has been done at the nation's research universities. Basic research has played a major role in technological breakthroughs in semiconductors, artificial intelligence, quantum, and biotechnology.

Federal funding for basic research at independent research universities is an economic security lever that supports U.S. national security. The deep cuts in federal funding for basic research in the current federal budget, along with attacks on the independence of those universities, are impairing this lever.

—Laura D'Andrea Tyson

ENDNOTES

1. Albert O. Hirschman, *National Power and the Structure of Foreign Trade*, 1st ed. (University of California Press, 1980), https://doi.org/10.2307/jj.15976659.

2. Henry Farrell and Abraham Newman, "The Weaponized World Economy," *Foreign Affairs*, August 19, 2025, https://www.foreignaffairs.com/united-states/weaponized-world-economy-farrell-newman.

3. Daniel W. Drezner, "How Everything Became National Security," *Foreign Affairs*, August 12, 2024, https://www.foreignaffairs.com/united-states/how-everything-became-national-security-drezner.

4. Japan passed an Economic Security Protection Act in 2022 and has established new government functions in this area. The European Union released an economic security strategy in early 2024 and is taking relevant action along a number of fronts. At the 2023 G7 summit in Hiroshima, G7 leaders issued a first-ever statement on economic security; Matthew P. Goodman, "G7 Gives First Definition to 'Economic Security,'" Center for Strategic and International Studies, May 31, 2023, https://www.csis.org/analysis/g7-gives-first-definition-economic-security.

5. Daleep Singh, "The Right Way to Wield America's Economic Power," *Foreign Affairs*, July 15, 2025, https://www.foreignaffairs.com/north-america/right-way-wield-americas-economic-power.

6. To be sure, there are other technologies that are important to economic security and warrant more attention, such as fusion, robotics, and digital currencies. The Task Force focused on AI, quantum, and biotech primarily because of their combined potential. They are also at different stages of development and face a diverse set of challenges that can offer broader lessons for the practice of economic security.

7. Michael Chiu et al., "The Economic Potential of Generative AI: The Next Productivity Frontier," McKinsey & Co., June 14, 2024, https://www.mckinsey.com/capabilities/mckinsey-digital/our-insights/the-economic-potential-of-generative-AI-the-next-productivity-frontier#business-value; Henning Soller, "The Year of Quantum: From Concept to Reality in 2025," McKinsey & Co., June 23, 2025, https://www.mckinsey.com/capabilities/mckinsey-digital/our-insights/the-year-of-quantum-from-concept-to-reality-in-2025/; and Anna Littmann, Mark Patel, Roberto Uchoa de Paula, and Pol van der Pluijm, "The Economic and Environmental Benefits of Advanced Biotechnology," McKinsey & Co., March 11, 2025, https://www.mckinsey.com/capabilities/sustainability/our-insights/sustainability-blog/the-economic-and-environmental-benefits-of-advanced-biotechnology.

8. Edward Fishman, *Chokepoints: American Power in the Age of Economic Warfare* (New York: Penguin Random House, 2025), https://www.penguinrandomhouse.com/books/726149/chokepoints-by-edward-fishman/.

9. Data reflects high-level analysis of recent investments, drawing on McKinsey & Company research.

10. Peter E. Harrell, "How to China-Proof the Global Economy," *Foreign Affairs*, December 12, 2023, https://www.foreignaffairs.com/china/how-china-proof-global-economy-america.

11. "Xi Mulls New Made-in-China Plan Despite U.S. Call to Rebalance," Bloomberg News, May 26, 2025, https://www.bloomberg.com/news/articles/2025-05-26/xi-plans-new-made-in-china-effort-even-as-trump-aims-to-boost-us-manufacturing; and "U.S. Efforts to Contain Xi's Push for Tech Supremacy Are Faltering," Bloomberg News, October 30, 2024, https://www.bloomberg.com/graphics/2024-us-china-containment/.

12. Jonathan Hillman, *The Digital Silk Road: China's Quest to Wire the World and Win the Future* (New York: Harper Business, 2021), https://www.harpercollins.com/products/the-digital-silk-road-jonathan-e-hillman?variant=41042005524514; and Sameer Patil and Prithvi Gupta, "The Digital Silk Road in the Indo-Pacific: Mapping China's Vision for Global Tech Expansion," Observer Research Foundation, January 3, 2024, https://www.orfonline.org/research/the-digital-silk-road-in-the-indo-pacific-mapping-china-s-vision-for-global-tech-expansion#_edn.

13. "Xi Jinping: Speech at the National Science and Technology Conference, the National Science and Technology Awards Conference, and the Conference of Academicians of the Chinese Academy of Sciences and the Chinese Academy of Engineering," Xinhua News Agency, June 24, 2024, http://www.news.cn/politics/leaders/20240624/16741a201e564d8d8775ffb1450ecf29/c.html. Translated by Manoj Kewalramani, "Breakdown of Xi's Speech on China's Quest for Science & Technology Strength," *Tracking People's Daily*, June 24, 2024, https://trackingpeoplesdaily.substack.com/p/breakdown-of-xis-speech-on-chinas.

14. Lennart Heim, "China's AI Models Are Closing the Gap—But America's Real Advantage Lies Elsewhere," RAND, May 2, 2025, https://www.rand.org/pubs/commentary/2025/05/chinas-ai-models-are-closing-the-gap-but-americas-real.html.

15. Nestor Maslej et al., *The AI Index 2025 Annual Report* (AI Index Steering Committee, Institute for Human-Centered AI, Stanford University, April 2025), https://hai.stanford.edu/assets/files/hai_ai_index_report_2025.pdf.

16. Hao-Ze Chen et al., "Implementation of Carrier-Grade Quantum Communication Networks over 10000 km," *npj Quantum Information* 11, no. 137 (2025), https://www.nature.com/articles/s41534-025-01089-8; Elizabeth Gibney, "Mini-Satellite Paves the Way for Quantum Messaging Anywhere on Earth," *Nature*, March 19, 2025, https://www.nature.com/articles/d41586-025-00581-7; and Sam Howell, *The Quest for Qubits: Assessing U.S.-China Competition in Quantum Computing* (Washington, DC: Center for a New American Security, May 2024), https://www.cnas.org/publications/reports/the-quest-for-qubits.

17. "China to Set Up National Venture Capital Guidance Fund, State Planner Says," Reuters, March 6, 2025, https://www.reuters.com/world/china/china-set-up-national-venture-capital-guidance-fund-state-planner-says-2025-03-06/.

18. Richard Baldwin, Rebecca Freeman, and Angelos Theodorakopoulos, "Hidden Exposure: Measuring U.S. Supply Chain Resilience," *Brookings Papers on Economic Activity*, September 27, 2023, https://www.brookings.edu/articles/hidden-exposure-measuring-us-supply-chain-reliance/.

19. Jiayu Zhang, "China's Military Employment of Artificial Intelligence and Its Security Implications," *International Affairs Review*, August 16, 2020, https://www.iar-gwu.org/print-archive/blog-post-title-four-xgtap.

20. "Winning the Race: America's AI Action Plan," White House, July 2025, https://www.whitehouse.gov/wp-content/uploads/2025/07/Americas-AI-Action-Plan.pdf.

21. Eric Rosenbach et al., *Critical and Emerging Technologies Index* (Cambridge, MA: Harvard Belfer Center, June 2025), https://www.belfercenter.org/critical-emerging-tech-index.

22. Nestor Maslej et al., "Chapter 4: Economy," in *The AI Index 2025 Annual Report*.

23. Ibid.; and Jacob Larson, James S. Denford, Gregory S. Dawson, and Kevin C. Desouza, "The Evolution of Artificial Intelligence (AI) Spending by the U.S. Government," Brookings Institution, March 26, 2024, https://www.brookings.edu/articles/the-evolution-of-artificial-intelligence-ai-spending-by-the-u-s-government/.

24. Xinmei Shen, "China's AI Capital Spending Set to Reach up to U.S. $98 Billion in 2025 Amid Rivalry With U.S.," *South China Morning Post*, June 25, 2025, https://www.scmp.com/tech/tech-war/article/3315805/chinas-ai-capital-spending-set-reach-us98-billion-2025-amid-rivalry-us.

25. Adam Barth, Chhavi Arora, Gayatri Shenai, Jesse Noffsinger, and Pankaj Sachdeva, "The Data Center Balance: How U.S. States Can Navigate the Opportunities and Challenges," McKinsey & Co., August 8, 2025, https://www.mckinsey.com/industries/public-sector/our-insights/the-data-center-balance-how-us-states-can-navigate-the-opportunities-and-challenges.

26. Alastair Green, Humayun Tai, Jesse Noffsinger, and Pankaj Sachdeva, "How Data Centers and the Energy Sector Can Sate AI's Hunger for Power," McKinsey & Co., September 17, 2024, https://www.mckinsey.com/industries/private-capital/our-insights/how-data-centers-and-the-energy-sector-can-sate-ais-hunger-for-power; and Jan Yong, "BRIGHTRAY Breaks Record by Completing Data Center in 8 Months," W.Media, May 28, 2025, https://w.media/brightray-breaks-record-by-completing-data-center-in-8-months/.

27. Mark Dangelo, "Needed AI Skills Facing Unknown Regulations and Advancements," Thomson Reuters, December 6, 2023, https://www.thomsonreuters.com/en-us/posts/technology/needed-ai-skills/.

28. Remco Zwetsloot and Jack Corrigan, *AI Faculty Shortages: Are U.S. Universities Meeting the Growing Demand for AI Skills?* (Washington, DC: Georgetown Center for Security and Emerging Technology, July 2022), https://cset.georgetown.edu/wp-content/uploads/CSET-AI-Faculty-Shortages.pdf.

29. "Over 700 Million People to Use AI Tools by 2030, Twice More Than This Year," AltIndex, April 1, 2025, https://altindex.com/news/million-people-use-ai-growth.

30. Radha Iyengar Plumb and Michael C. Horowitz, "What America Gets Wrong About the AI Race," *Foreign Affairs*, April 18, 2025, https://www.foreignaffairs.com/united-states/what-america-gets-wrong-about-ai-race.

31. "Winning the Race," White House.

32. "China Issues Guideline to Accelerate 'AI Plus' Integration Across Key Sectors," State Council of the People's Republic of China, originally posted by ChinaDaily.Com, August 27, 2025, https://english.www.gov.cn/policies/latestreleases/202508/27/content_WS68ae7976c6d0868f4e8f51a0.html; Stephen Chen, "How AI Revolution Is Making a Chinese Coal Mine Turn More Profits Than an Investment Bank," *South China Morning Post*, March 31, 2025, https://www.scmp.com/news/china/science/article/3304115/how-ai-revolution-making-chinese-coal-mine-turn-more-profits-investment-bank; and Hodan Omaar, "How Innovative Is China in AI?," Information Technology & Innovation Foundation, August 26, 2024, https://itif.org/publications/2024/08/26/how-innovative-is-china-in-ai/.

33. Raj Varadarajan et al., *Emerging Resilience in the Semiconductor Supply Chain* (Semiconductor Industry Association, May 2024), https://www.semiconductors.org/wp-content/uploads/2024/05/Report_Emerging-Resilience-in-the-Semiconductor-Supply-Chain.pdf.

34. Saloni Gankar, *Industrial Semiconductor Market Tracker – 3Q22 Analysis*, Omdia, February 21, 2023, https://omdia.tech.informa.com/om028963/industrial-semiconductor-market-tracker----3q22-analysis.

35. Servicemember Quality of Life Improvement and National Defense Authorization Act for Fiscal Year 2025, Pub. L. No. 118–159, 138 Stat. 1773 (2024), https://www.congress.gov/bill/118th-congress/house-bill/5009/text.

36. Eric Pogue et al., "Understanding Current Battery and BESS Supply Chain Risks," Willkie Farr & Gallagher LLP, December 11, 2024, https://www.willkie.com/-/media/files/publications/2024/12/understanding-current-battery-and-bess-supply-chain-risks.pdf.

37. "Optical Transceivers for Datacom and Telecom 2024," YOLE Group, May 2024, https://www.yolegroup.com/product/report/optical-transceivers-for-datacom-2024/.

38. U.S. Geological Survey, *Mineral Commodity Summaries 2025* (St. Louis: U.S. Government Publishing Office, March 2025), https://doi.org/10.3133/mcs2025; and Keith Bradsher, "China Halts Critical Exports as Trade War Intensifies," *New York Times*, April 13, 2025, https://www.nytimes.com/2025/04/13/business/china-rare-earths-exports.html.

39. For example, export-controlled chips like Nvidia's H100 and A100 have made their way to mainland China through shipments from shell companies based overseas that are relabeled as other goods, falsified end-user reports from buyers based in Singapore, and human smuggling around Southeast Asia. See Ana Swanson and Claire Fu, "With Smugglers and Front Companies, China Is Skirting American AI Bans," *New York Times*, August 4, 2024, https://www.nytimes.com/2024/08/04/technology/china-ai-microchips.html; Xinghui Kok, "Singapore Charges Three With Fraud That Media Link to Nvidia Chips," Reuters, February 28, 2025, https://www.reuters.com/technology/singapore-charges-three-with-fraud-that-media-link-nvidia-chips-2025-02-28/; and Nadine Chua, "New Charges for 3 Men Allegedly Linked to Servers That Likely Contained Nvidia Chips," *Straits Times*, March 6, 20205, https://www.straitstimes.com/singapore/courts-crime/new-charges-for-2-of-the-3-men-allegedly-linked-to-computer-servers-that-likely-contained-nvidia.

40. Matthew S. Axelrod, "Assistant Secretary for Export Enforcement Matthew S. Axelrod Delivers Remarks at the Practising Law Institute's Coping with U.S. Export Controls and Sanctions Conference," remarks as prepared for delivery, December 9, 2024, https://www.bis.gov/speech/assistant-secretary-export-enforcement-matthew-s.-axelrod-delivers-remarks-practising-law-institutes-coping-u.s.

41. Dylan Tokar, "Seagate to Pay $300 Million for Violating Export Restrictions on China's Huawei," *Wall Street Journal*, April 19, 2023, https://www.wsj.com/articles/seagate-to-pay-300-million-for-violating-export-restrictions-on-chinas-huawei-9fbdc1a.

42. "Winning the Race," White House.

43. "Quantum Technology Monitor," McKinsey Digital, April 2024, https://www.mckinsey.com/~/media/mckinsey/business%20functions/mckinsey%20digital/our%20insights/steady%20progress%20in%20approaching%20the%20quantum%20advantage/quantum-technology-monitor-april-2024.pdf.

44. Soller, "The Year of Quantum."

45. Nick Vonyhady, "Labor's Bold $1b Bet on Aussie Quantum Start-Up," AFR, April 29, 2024, https://www.afr.com/technology/labor-s-bold-1b-bet-on-aussie-quantum-start-up-20240429-p5fnb7; Illinois.gov, "Gov Pritzker Announces Location and PsiQuantum as Anchor Tenant of New Quantum Park," press release, https://www.illinois.gov/news/release.html?releaseid=30472; and University of Illinois System, "DARPA, Gov. Pritzker Pledge Funding for Quantum Proving Ground," press release, July 17, 2024, http://news.uillinois.edu/view/7815/1344185859.

46. Zhang Zhihao, "Xi Highlights Crucial Role of Quantum Tech," *ChinaDaily*, October 19, 2020, https://www.chinadailyhk.com/hk/article/189362.

47. "Quantum Technology Monitor," McKinsey Digital.

48. "Quantum Key Distribution (QKD) and Quantum Cryptography (QC)," National Security Agency/Central Security Service, n.d., https://www.nsa.gov/Cybersecurity/Quantum-Key-Distribution-QKD-and-Quantum-Cryptography-QC/#:~:text=Quantum%20key%20distribution%20utilizes%20the,updated%20guidance%20through%20CNSSP%2D15.

49. Matt Swayne, "China Establishes Quantum-Secure Communication Links With South Africa," Quantum Insider, March 14, 2025, https://thequantuminsider.com/2025/03/14/china-established-quantum-secure-communication-links-with-south-africa/.

50. "Is China a Leader in Quantum Technologies?," *ChinaPower*, Center for Strategic and International Studies, August 14, 2023, https://chinapower.csis.org/china-quantum-technology/.

51. Stephen Chen, "China Launches World's Fastest Programmable Quantum Computers," *South China Morning Post*, October 26, 2021, https://www.scmp.com/news/china/science/article/3153727/china-launches-worlds-fastest-programmable-quantum-computers; Frank Arute et al., "Quantum Supremacy Using a Programmable Superconducting Processor," *Nature* 574, no. 505–510 (2019), https://www.nature.com/articles/s41586-019-1666-5; "China Becomes 3rd Country to Deliver Quantum Computers, after Canada and U.S.," *Global Times*, January 30, 2023, https://www.globaltimes.cn/page/202301/1284467.shtml; Antonia Hmaidi and Jeroen Groenewegen-Lau, "China's Long View on Quantum Tech Has the U.S. and EU Playing Catch-Up," Mercator Institute for China Studies, December 12, 2024, https://merics.org/en/report/chinas-long-view-quantum-tech-has-us-and-eu-playing-catch; and Matt Swayne, "Chinese Team Officially Report on Zuchongzhi 3.0, Claims Million Times Speedup Over Google's Sycamore," Quantum Insider, May 9, 2025, https://thequantuminsider.com/2025/03/04/chinese-team-officially-report-on-zuchongzhi-3-0-claims-million-times-speedup-over-googles-willow/.

52. Celia Merzbacher, "Addressing the U.S. Quantum Labor Shortage Before It's Too Late," *MeriTalk*, July 23, 2024, https://www.meritalk.com/addressing-the-u-s-quantum-labor-shortage-before-its-too-late.

53. "The Quantum Insider Projects $1 Trillion in Economic Impact From Quantum Computing by 2035," Quantum Insider, September 13, 2024, https://thequantuminsider.com/2024/09/13/the-quantum-insider-projects-1-trillion-in-economic-impact-from-quantum-computing-by-2035/.

54. "Commerce Control List Additions and Revisions; Implementation of Controls on Advanced Technologies Consistent With Controls Implemented by International Partners," Federal Register, September 6, 2024, https://www.federalregister.gov/documents/2024/09/06/2024-19633/commerce-control-list-additions-and-revisions-implementation-of-controls-on-advanced-technologies.

55. Brad Smith, "Investing in American Leadership in Quantum Technology: The Next Frontier in Innovation," Microsoft, April 28, 2025, https://blogs.microsoft.com/on-the-issues/2025/04/28/investing-in-american-leadership-quantum/.

56. Subcommittee on Quantum Information Science, *Advancing International Cooperation in Quantum Information Science and Technology* (Washington, DC: National Science and Technology Council, August 2024), https://www.quantum.gov/wp-content/uploads/2024/08/Advancing-International-Cooperation-in-QIST.pdf.

57. Foreign countries of concern are commonly defined to include China, Iran, North Korea, and Russia—under laws like the CHIPS and Science Act and related export control regimes.

58. National Security Commission on Emerging Biotechnology, *Charting the Future of Biotechnology* (Washington, DC: National Security Commission on Emerging Biotechnology, April 2025), https://www.biotech.senate.gov/final-report/chapters/.

59. Greg Licholai, "Global Drug Development Shifts East," *Forbes*, June 25, 2025, https://www.forbes.com/sites/greglicholai/2025/06/25/global-drug-development-shifts-east/.

60. Chris Bradley et al., "The Next Big Arenas of Competition," McKinsey Global Institute, October 23, 2024, https://www.mckinsey.com/mgi/our-research/the-next-big-arenas-of-competition.

61. Rosenbach et al., *Critical and Emerging Technologies Index*.

62. "Labor Force Statistics from the Current Population Survey," U.S. Bureau of Labor Statistics, last updated January 29, 2025, https://www.bls.gov/cps/cpsaat11b.htm.

63. Stephanie A. Freel et al., "Now Is the Time to Fix the Clinical Research Workforce Crisis," *Clinical Trials* 20, no. 5 (2023): 457–62, https://pmc.ncbi.nlm.nih.gov/articles/PMC10504806/.

64. Merck, "Merck Enters Exclusive License Agreement for HRS-5346," press release, March 25, 2025, https://www.merck.com/news/merck-enters-exclusive-license-agreement-for-hrs-5346-an-investigational-oral-lipoproteina-inhibitor-for-cardiovascular-disease-from-jiangsu-hengrui-pharmaceuticals-co-ltd/; and Pfizer, "Pfizer Enters Into Exclusive Licensing Agreement with 3SBio," press release, May 19, 2025, https://www.pfizer.com/news/press-release/press-release-detail/pfizer-enters-exclusive-licensing-agreement-3sbio.

65. Andrew Rechenberg, "U.S. Dangerously Reliant on High-Risk Imported Drug Supply," Coalition for a Prosperous America, May 29, 2025, https://prosperousamerica.org/u-s-dangerously-reliant-on-high-risk-imported-drug-supply/.

66. "The Pharmaceutical Industry: An Overview of CPI, PPI, and IPP Methodology," U.S. Bureau of Labor Statistics, December 2021, https://www.bls.gov/ppi/methodology-reports/the-pharmaceutical-industry-an-overview-of-cpi-ppi-and-ipp-methodology.pdf; and Anthony Sardella, "The U.S. Active Pharmaceutical Ingredient Infrastructure: The Current State and Considerations to Increase U.S. Healthcare Security," Washington University Center for Analytics and Business Insights, August 1, 2021, https://wustl.app.box.com/s/rjo1i7yews99hdr8zeo5fp0u71g47m0i.

67. Marta E. Wosińska and Yihan Si, "U.S. Drug Supply Chain Exposure to China," Brookings Institution, July 28, 2025, https://www.brookings.edu/articles/us-drug-supply-chain-exposure-to-china/.

68. Anthony Sardella, "The U.S. Active Pharmaceutical Ingredient Infrastructure: The Current State and Considerations to Increase U.S. Healthcare Security"; and Vic Suarez, "The National Security Rationale for Stockpiling Key Pharmaceutical Ingredients," Council on Strategic Risks Janne E. Nolan Center, March 5, 2024, https://councilonstrategicrisks.org/wp-content/uploads/2024/03/57-Stockpiling-Pharma-Ingredients.pdf.

69. Scott Gottlieb, "How To Stop the Shift of Drug Discovery From the U.S. to China," *STAT*, May 6, 2025, https://www.statnews.com/2025/05/06/how-to-stop-the-shift-of-drug-discovery-from-the-u-s-to-china/.

70. Anthony Sardella, "The U.S. Active Pharmaceutical Ingredient Infrastructure: The Current State and Considerations to Increase U.S. Healthcare Security."

71. Outsourcing is occurring most significantly through contract research organizations and contract development manufacturing organizations.

72. "Trade Association Survey Shows 79% of U.S. Biotech Companies Contract with Chinese Firms," Reuters, May 8, 2024, https://www.reuters.com/business/healthcare-pharmaceuticals/trade-association-survey-shows-79-us-biotech-companies-contract-with-chinese-2024-05-08/.

73. Christina Jewett, "Chinese Company Under Congressional Scrutiny Makes Key U.S. Drugs," *New York Times*, April 15, 2024, https://www.nytimes.com/2024/04/15/health/wuxi-us-drugs-congress.html.

74. 170 Cong. Rec. H50501 (Sept. 9, 2024), https://www.congress.gov/118/crec/2024/09/09/170/139/CREC-2024-09-09-pt1-PgH5051.pdf.

75. J. Christopher Love, Elisabeth B. Reynolds, David Goldston, and Hannah E. Frye, "Biomanufacturing in the U.S.: A MIT Policy Brief," Massachusetts Institute of Technology, January 30, 2025, https://dspace.mit.edu/handle/1721.1/158134.

76. National Security Commission on Emerging Biotechnology, "Section 2.2: Attract and Scale Private Capital to Support Biotechnology," in *Charting the Future of Biotechnology*, https://www.biotech.senate.gov/final-report/chapters/chapter-2/section-2/#rec-2-2a.

77. "G7 Critical Minerals Action Plan," G7 2025, June 17, 2025, https://g7.canada.ca/en/news-and-media/news/g7-critical-minerals-action-plan/.

78. U.S. Departments of Labor, Commerce, and Education, *America's Talent Strategy: Building*

the Workforce for the Golden Age (Washington, DC: U.S. Departments of Labor, Commerce, and Education, August 2025), https://www.dol.gov/sites/dolgov/files/OPA/newsreleases/2025/08/Americas-Talent-Strategy-Building-the-Workforce-for-the-Golden-Age.pdf.

79. Council on Foreign Relations, *The Work Ahead: Machines, Skills, and U.S. Leadership in the Twenty-First Century* (New York: Council on Foreign Relations, 2018), https://www.cfr.org/task-force-report/work-ahead.

80. Jonathan Chanis, "The 47th Presidency Offers Hope for Permitting Reform," Utility Dive, January 16, 2025, https://www.utilitydive.com/news/Trump-hope-energy-permitting-reform-transmission-renewables/737507/.

81. In 2018, Congress substantially updated foreign investment screening through the Foreign Investment Risk Review Modernization Act. Outbound investment screening has been in place for less than a year. Much of the machinery behind U.S. sanctions was updated after the September 11, 2001, terrorist attacks, with notable innovations in response to Russia's invasion of Ukraine, and is worthy of further attention as the primary challenge has shifted away from nonstate actors to great powers.

82. William Henagan, "Reauthorizing DFC: A Primer for Policymakers," Council on Foreign Relations, March 31, 2025, https://www.cfr.org/article/reauthorizing-dfc-primer-policymakers.

83. Daleep Singh, "The Right Way to Wield America's Economic Power."

ACRONYMS

AI
artificial intelligence

API
active pharmaceutical ingredients

ARPA-E
Advanced Research Projects Agency-Energy

AUKUS
the trilateral security agreement between Australia, the United Kingdom, and the United States

BESS
battery energy storage systems

BIS
Bureau of Industry and Security

CDMO
contract development and manufacturing organization

CFIUS
Committee on Foreign Investment in the United States

CRO
contract research organization

DARPA
Defense Advanced Research Projects Agency

DFC
Development Finance Corporation

DOD
Department of Defense

DOE
Department of Energy

EU
European Union

IC
integrated circuit

ICTS
information and communications technology and services

IP
intellectual property

IT
information technology

KSM
key starting materials

NDS
National Defense Stockpile

NIST
National Institute of Standards and Technology

NQI
National Quantum Initiative

NSF
National Science Foundation

PCB
printed circuit board

PCBA
printed circuit board assembly

PLA
People's Liberation Army

QIS
quantum information science

R&D
research and development

STEM
science, technology, engineering, and mathematics

ACKNOWLEDGMENTS

This report reflects the dedication of the members and observers of the Task Force on Economic Security, an accomplished group who devoted months to grappling with these complex challenges. I am especially grateful to our distinguished cochairs, Gina Raimondo, Justin Muzinich, and James Taiclet, for their steady leadership and focus on delivering practical recommendations.

We benefited from many candid, off-the-record conversations with leaders in government and business. Former officials helped shape our principles for practicing economic security, and business leaders shared their frontline views of today's competition for foundational technologies. The staff of the Senate National Security Commission on Emerging Biotechnology sharpened our recommendations on biotech. A talented McKinsey team—including Jon Spaner, Ziad Haider, Bill Wiseman, Henning Soler, Alex Singla, Sara O'Rourke, Mary Parker Aldecoa, Caitlyn Mason, James Sorensen, Andrew Kennedy, and Charlotte Rediker—offered insights into supply chain risks and private capital mobilization. We also drew on thoughtful contributions from CFR members who joined discussions in New York, Washington, DC, and beyond. I would also like to thank the Amy Falls and Hartley Rogers Foundation for its generous support of the RealEcon Initiative.

This project would not have been possible without the extraordinary support of my CFR colleagues. I would like to thank Shannon O'Neil, Matt Goodman, Tom Bollyky, Heidi Crebo-Rediker, Rush Doshi, Kat Duffy, Jessica Harrington, William Henagan, Mike Horowitz, Rebecca Patterson, Stuart Reid, Adam Segal, Chris Tuttle, and Laura Taylor-Kale. My research associate Ishaan Thakker and intern Liam Hurtt provided critical research support and creative solutions as we moved from idea to outline and from draft to draft. The Product, Design, and Publications teams, steered by Maria Teresa Alzuru, expertly produced this report, including graphics by Will Merrow that help clarify complex issues. The Task Force team, led by Anya Schmemann, and assisted by Chelie Setzer and Katerina Viyella, embraced new approaches and kept this project moving forward at every step. Finally, I am grateful to CFR President Mike Froman for being one of the first practitioners and thought leaders to recognize how economic power was taking on a more central role in U.S. national security, and for the opportunity to direct this Task Force.

Jonathan E. Hillman
Project Director

TASK FORCE MEMBERS

Task Force members are asked to join a consensus signifying that they endorse "the general policy thrust and judgments reached by the group, though not necessarily every finding and recommendation." They participate in the Task Force in their individual, not institutional, capacities.

Dr. Noubar Afeyan is founder and CEO of Flagship Pioneering, a company that creates bioplatform companies to transform human health and sustainability. An entrepreneur and biochemical engineer, Afeyan holds more than one hundred patents and has cofounded more than one hundred life science and technology start-ups during his thirty-seven-year career. He is cofounder and chairman of the board of Moderna, the pioneering messenger RNA company. He was a senior lecturer at MIT's Sloan School of Management and a lecturer at Harvard Business School, and he currently serves as a member of the MIT Corporation. He is a member of the Council on Foreign Relations and was elected to the National Academy of Engineering in 2022. Afeyan is a trustee of the Afeyan Foundation and a cofounder of the Aurora Humanitarian Initiative, as well as several other philanthropic projects. Afeyan completed his undergraduate studies at McGill University and his doctoral work in biochemical engineering at MIT.

Kevin Brown is currently an executive vice president and chief supply chain officer at Dell Technologies. He has over thirty years of leadership experience in operations, technology, and procurement. In his two decades of leadership at Dell, the company has been recognized as having one of the most efficient, sustainable, and innovative supply chains across industries worldwide. Prior to joining Dell, Brown spent ten years at Newport News Shipbuilding in various leadership roles in engineering, construction management, and facilities. Brown serves on Kroger's board of directors, as well as the boards of the John F. Kennedy Library Foundation and the Howard University Center for Supply Chain Excellence. He is also a member of the George Washington University board of trustees and is a life member of the Council on Foreign Relations. Brown holds a BS in mechanical engineering from the University of Massachusetts at Amherst and an MS in engineering management from the George Washington University.

Martin Chorzempa is a senior fellow at the Peterson Institute for International Economics. His research focuses on financial technology and digital currency, as well as technology and national security issues like export controls, foreign investment screening, and artificial intelligence. He is author of *The Cashless Revolution: China's Reinvention of Money*, which the *Financial Times* named one of the best economics books of 2022. He is regularly quoted by major media outlets, including the *Wall Street Journal*, *New York Times*, *Washington Post*, *Financial Times*, and *Foreign Affairs*. He has been a Fulbright Scholar in Germany and a Luce Scholar at Peking University's China Center for Economic Research, and he worked for the China Finance 40 Forum, a leading independent think tank, in Beijing. In 2017, he graduated from the Harvard Kennedy School with a master's in public administration in international development.

Sarah Bauerle Danzman is an associate professor of international studies at Indiana University, Bloomington's Hamilton Lugar School of Global and International Studies and director of the Indiana University Tobias Center for Innovation in International Development. She is a nonresident senior fellow at the Atlantic Council and a term member of the Council on Foreign Relations. From 2019 to 2020, she was a CFR international affairs fellow, working in the U.S. Department of State as a policy advisor in the Office of Investment Affairs, where she staffed the Committee on Foreign Investment in the United States. Bauerle Danzman is an editor in chief of the Brill journal *Law & Geoeconomics*. She is the author of *Merging Interests: When Domestic Firms Shape FDI Policy* and many articles published in scholarly and policy outlets, including *Foreign Affairs, International Studies Quarterly, Review of International Political Economy*, and the *European Journal of International Relations*. Her research explores how domestic and multinational firms influence and adapt to investment regulation, the nexus of national security and investment, and how rules governing capital shape global networks of ownership, production, and economic growth.

Thomas Donilon is vice chairman of BlackRock. He is also chairman of the BlackRock Investment Institute. Previously, Donilon served as U.S. national security advisor to President Barack Obama. In that capacity, he was responsible for the coordination and integration of the administration's foreign policy, intelligence, and military efforts and also served as the president's emissary to a number of world leaders. Donilon chaired the Obama-Biden transition at the U.S. State Department and National Security Council and headed President Obama's debate preparation during the 2008 campaign. He has also served as chair of the Presidential Commission to Enhance National Cybersecurity, cochair of the Foreign Policy Advisory Board at the U.S. State Department, and for eight years as a member of the Defense Policy Board and the CIA External Advisory Board. He is a member of the Board of Directors of the Council on Foreign Relations. Donilon has worked with and advised four U.S. presidents since his first position at the White House in 1977, working with President Jimmy Carter. Donilon has received the Secretary of State's Distinguished Service Award, the National Intelligence Distinguished Public Service Medal, the Department of Defense Medal for Distinguished Public Service, the Chairman of the Joint Chiefs of Staff Joint Distinguished Civilian Service Award, and the CIA's Director's Award.

Elizabeth C. Economy is the Hargrove senior fellow and codirector of the US, China, and the World Program at Stanford University's Hoover Institution. From 2021 to 2023, Economy served as the senior advisor for China to the secretary of commerce. Economy is an acclaimed expert on Chinese domestic and foreign policy and U.S.-China relations. An award-winning author, her most recent book is *The World According to China*, published in 2021. She is also the author of *The Third Revolution: Xi Jinping and the New Chinese State; By All Means Necessary: How China's Resource Quest Is Changing the World*, with Michael Levi; and *The River Runs Black: The Environmental Challenge to China's Future*. Her books have been translated into a dozen languages. Economy has published opinion pieces in the *Washington Post*, *Wall Street Journal*, *Financial Times*, and *New York Times*, and appears frequently on nationally broadcast television programs, including *Fareed Zakaria GPS*, *PBS NewsHour*, and Bloomberg's *Wall Street Week*. She serves on the boards of Swarthmore College, the National Endowment for Democracy, and the National Committee on U.S.-China Relations and is a member of the Aspen Strategy Group and Council on Foreign Relations. She also serves as the East Asia book reviewer for *Foreign Affairs*.

Edward Fishman is a senior research scholar at the Center on Global Energy Policy and an adjunct professor at Columbia University, where he teaches courses on economic statecraft and geoeconomics. He is the *New York Times* best-selling author of *Chokepoints: American Power in the Age of Economic Warfare*. He also advises companies on geopolitical strategy and invests in early-stage technology start-ups. Previously, he served at the U.S. State Department as a member of the secretary of state's policy planning staff and as Russia and Europe sanctions lead, at the Pentagon as an advisor to the chairman of the Joint Chiefs of Staff, and at the U.S. Treasury Department as special assistant to the undersecretary for terrorism and financial intelligence. In the private sector, Fishman has worked in operating roles at several high-growth technology companies. His writing and analysis are regularly featured by outlets such as the *New York Times*, *Wall Street Journal*, *Washington Post*, *Foreign Affairs*, Politico, and NPR. He holds a BA in history from Yale, an MPhil in international relations from Cambridge, and an MBA from Stanford.

Michèle A. Flournoy is cofounder and managing partner of WestExec Advisors, and a cofounder, former chief executive officer, and now chair of the Center for a New American Security (CNAS). Flournoy served as the undersecretary of defense for policy from February 2009 to February 2012. She was the principal advisor to the secretary of defense in the formulation of national security and defense policy, oversight of military plans and operations, and in National Security Council deliberations. She led the development of the Department of Defense's 2012 Strategic Guidance and represented the department in foreign engagements, media, and before Congress. Prior to confirmation, Flournoy coled President Obama's transition team at the Defense Department. In 2007, she cofounded CNAS, a bipartisan think tank dedicated to developing pragmatic and principled national security policies. She served as president of CNAS until 2009 and returned as CEO in 2014. In 2017, she cofounded WestExec Advisors, a strategic advisory firm. Flournoy serves on the boards of CNAS, Booz Allen Hamilton, the Council on Foreign Relations, Amida Technology Solutions, and CARE, as well as serving on a number of geopolitics– and national security–focused advisory boards. Flournoy earned a bachelor's degree from Harvard University and a master's degree from Balliol College, Oxford University.

Jonathan E. Hillman is senior fellow for geoeconomics at the Council on Foreign Relations. His expertise spans economic and security issues, including investment, trade, infrastructure, and technology. He is the author of *The Digital Silk Road: China's Quest to Wire the World and Win the Future* and *The Emperor's New Road: China and the Project of the Century*. Hillman has served as a senior advisor to three U.S. cabinet officials. From 2023 to 2024, he advised the secretary of commerce on finance, infrastructure, and trade issues. From 2022 to 2023, he served on the secretary of state's policy planning staff as the lead member covering economics and colead for technology. From 2014 to 2016, he directed the research and writing process at the office of the U.S. trade representative for reports, speeches, and other materials explaining U.S. trade and investment policy. From 2016 to 2022, Hillman was a senior fellow and director of the Reconnecting Asia Project at the Center for Strategic and International Studies. He has worked as a researcher at the Belfer Center for Science and International Affairs, at the Council on Foreign Relations, and in Kyrgyzstan as a Fulbright Scholar. Hillman is a graduate of the Harvard

Kennedy School, where he was a presidential scholar, and Brown University, where he was elected to Phi Beta Kappa and received the Garrison Prize for best thesis in international relations.

Karen Karniol-Tambour is cochief investment officer at Bridgewater Associates, responsible for managing the company's investment process. She oversees the systemization of Bridgewater's research into trading strategies, manages the development of proprietary investment management models, directs the design and implementation of client investment strategies, and publishes timely market understanding to clients and global policy makers via Bridgewater's Daily Observations. Karniol-Tambour also leads the firm's investing in Asia and China strategies. She joined Bridgewater in 2006 after graduating from Princeton University. Outside of Bridgewater, Karniol-Tambour is a World Economic Forum Young Global Leader, a term member of the Council on Foreign Relations, and serves on the Atlantic Council's board of directors. She is also a member of the Robin Hood Foundation Investment Committee and serves on the Financial Services Advisory Council for the Federal Reserve Bank of Dallas. As a committed advocate of enhancing women's leadership, she is an active angel investor focused on technology start-ups led by women. Karniol-Tambour has been recognized across the industry for her achievements, including *Fortune*'s 40 Under 40 most influential leaders in business in 2019, and she has been named to Barron's list of Most Influential Women in U.S. Finance for five years in a row.

Aditi Kumar is a senior fellow at Harvard Kennedy School's Belfer Center for Science and International Affairs and a nonresident fellow at the Center for Security and Emerging Technology. Kumar is the former principal deputy director of the Defense Innovation Unit (DIU) within the U.S. Department of Defense. In that role, she led numerous initiatives focused on the accelerated acquisition and fielding of commercial technologies in response to urgent military needs. Prior to joining DIU, Kumar served as the senior advisor to the undersecretary of defense for acquisition and sustainment, leading efforts related to industrial production expansion and acceleration. She was previously the executive director of the Belfer Center for Science and International Affairs. She also led the Belfer Center's Economic Diplomacy Initiative, focused on research and expertise at the intersection of international affairs and economic policy. Prior to her time at Harvard, she was a principal at management consultancy Oliver Wyman in the financial services and public policy practices. Kumar holds a BS in economics and a BA in international studies from the University of Pennsylvania's Huntsman Program, an MBA from the Harvard Business School, and a master's in public policy from the Harvard Kennedy School.

Sigal Mandelker joined Ribbit Capital in April 2020. Ribbit is an investment firm focused on financial services and technology. Prior to Ribbit, Mandelker served as acting deputy secretary of the U.S. Treasury Department and as undersecretary for terrorism and financial intelligence. As undersecretary, Mandelker supervised four main components of Treasury: the Office of Foreign Asset Controls, the Financial Crimes Enforcement Network, the Office of Intelligence and Analysis, and the Office of Terrorist Financing and Financial Crimes. Before serving at Treasury, Mandelker was a partner at Proskauer in New York. Mandelker also previously served in a number of senior positions in the U.S. government, including as deputy assistant attorney general in the Criminal Division of the Justice Department, an assistant U.S. attorney in the U.S. Attorney's Office for the Southern District of New York, counselor to Secretary of Homeland Security Michael Chertoff, counsel to the deputy attorney general, and as a law clerk on the U.S. Supreme Court. Mandelker is also an advisor to Chainalysis, has sat on the boards of the Crypto Council for Innovation and the Financial Technology Association, served as cochair of the Center for a New American Security's task force on fintech, crypto, and national security, and is also a member of the Council on Foreign Relations. She served as a member of the U.S. Holocaust Memorial Museum Council and executive committee, as well as chair of the museum's Collections and Acquisitions Committee.

Brent J. McIntosh is Citi's chief legal officer & corporate secretary, overseeing Citi's Global Legal Affairs & Compliance organization, which includes the Legal Department, Independent Compliance Risk Management, Citi Security and Investigative Services, and Citi's Regulatory Strategy and Policy function. Before joining Citi, McIntosh served as undersecretary of the treasury for international affairs from 2019 to 2021 and as Treasury's general counsel from 2017 to 2019, spearheading Treasury's regulatory reform efforts. He had previously been a partner at Sullivan & Cromwell. McIntosh served in the White House from 2006 until 2009, first as associate counsel to the president and then as deputy assistant to the president and deputy staff secretary. Before that, he was a deputy assistant attorney general at the Justice Department. A Michigan native, McIntosh holds an AB in economics and political science from the University of Michigan and a JD from Yale Law School.

Chris Miller is the author of *Chip War: The Fight for the World's Most Critical Technology*, revealing the geopolitical impact of the computer chip. It is a *New York Times* best seller and a winner of the *Financial Times* Business Book of the Year Award, the Council on Foreign Relations Arthur Ross Book Award, and the Institute of Electrical and Electronics Engineers 2024 History Prize.

Miller is professor at Tufts University's Fletcher School and is a nonresident senior fellow at the American Enterprise Institute. Miller is frequently featured in media such as the *New York Times*, *Wall Street Journal*, and *Financial Times*, as well as on NPR and CNBC. In addition to *Chip War*, his books include *We Shall Be Masters: Russian Pivots to East Asia from Peter the Great to Putin*, *Putinomics: Power and Money in Resurgent Russia*, and *The Struggle to Save the Soviet Economy: Mikhail Gorbachev and the Collapse of the USSR*. He has a BA in history from Harvard University and an MA and PhD in history from Yale University.

Craig Mundie is a distinguished technology executive with a career defined by significant contributions to the technology industry and public policy. He advises organizations and governments on emerging technologies, particularly in areas such as fusion energy, cybersecurity, quantum computing, molecular medicine, and artificial intelligence. Mundie spent twenty-two years at Microsoft, holding leadership roles including chief technical officer, chief research and strategy officer, and senior advisor to the CEO, where he shaped the company's technology strategy and research initiatives. He continues to advise Microsoft on strategic initiatives. In addition to his corporate achievements, Mundie served for ten years on the U.S. National Security Telecommunications Advisory Committee, offering critical and trusted guidance to three U.S. administrations on tech and national security. He was also part of President Obama's Council of Advisors on Science and Technology for eight years, helping shape national science and technology policies. Mundie is a leading advocate for responsible technology development, particularly in the fields of artificial intelligence, advanced computing systems, and global technology policy. He coauthored *Genesis: Artificial Intelligence, Hope, and the Human Spirit* with Henry Kissinger and Eric Schmidt. Mundie currently cochairs a Track II dialogue with China on artificial intelligence, continuing his work in global technology policy and innovation.

Justin G. Muzinich is CEO of Muzinich & Co., an investment firm. He served as the U.S. deputy secretary of the treasury from 2018 to 2021, with broad responsibility for U.S. economic policy. While deputy secretary, Muzinich was also responsible for the roles of undersecretary for terrorism and financial intelligence and undersecretary for domestic finance, managing the divisions of Treasury that oversee national security policy and financial policy. Muzinich played a leading role in the economic response to COVID-19 and represented the United States at the Group of Seven, Group of Twenty, and Organization of Economic Cooperation and Development meetings. His national security responsibilities included the Committee on Foreign Investment in the United

States (CFIUS) and a wide range of National Security Council matters. From 2017 to 2018, Muzinich served as counselor to the secretary of the treasury and helped lead the effort to reform the U.S. tax code. From 2015 to 2016, he was policy director for Jeb Bush's presidential campaign. After leaving public service, he served as a distinguished fellow at CFR and as a senior fellow at Harvard Kennedy School. Before public service, he worked in finance and taught at Columbia Business School. He is on the boards of New York-Presbyterian Hospital, the National Bureau of Economic Research, and the Council on Foreign Relations. He earned his AB from Harvard College, his JD from Yale Law School, and his MBA from Harvard Business School.

Nazak Nikakhtar is an international trade and national security attorney; a partner at the Washington, DC, law firm of Wiley Rein LLP; and cochair of Wiley's national security practice and CFIUS practice. She is also a trade and industry economist and a former Georgetown University adjunct law professor. From 2018 to 2021, with unanimous confirmation by the U.S. Senate, Nikakhtar served as the Department of Commerce's assistant secretary for industry and analysis at the International Trade Administration. Nikakhtar also served the duties of the undersecretary for industry and security at Commerce's Bureau of Industry and Security. As one of the key national security experts in the U.S. government, she developed and implemented innovative laws, regulations, and policies to safeguard strategically important technologies, strengthen the U.S. industrial base, and protect the national security and foreign policy interests of the United States. As the department's lead on CFIUS, she played a key role in shaping U.S. investment policy. As the head of the agency's trade policy office, she advised the U.S. government on legal and economic issues impacting critical technologies, advanced manufacturing, financial services, e-commerce, data privacy, cybersecurity, critical minerals and rare earths, and energy competition.

Gina M. Raimondo is a distinguished fellow at the Council on Foreign Relations. From March 2021 through January 2025, she was the U.S. secretary of commerce. As secretary, she focused on a vital mission—making the United States more competitive through good-paying jobs, empowering entrepreneurs to innovate, and advancing economic and national security. The Department of Commerce, under her leadership, made historic investments in internet access, manufacturing, economic development, workforce training, supply chain resiliency, and climate readiness through the Bipartisan Infrastructure Law, the CHIPS and Science Act, and the Inflation Reduction Act. Raimondo took the lead in ensuring the responsible development of artificial intelligence, standing up the U.S. AI Safety Institute in the Commerce

Department. During her service, the department released industry-standard guidance on red teaming, generative AI, and synthetic content; spearheaded federal efforts to mitigate national security threats posed by AI; and launched the international network of AI Safety Institutes. Previously, Raimondo was the seventy-fifth and first-woman governor of Rhode Island. As governor, she kick-started the state's economy, making record investments in infrastructure, education, and job training. She was reelected by the widest margin in a generation. Before serving as governor, Raimondo was the state's general treasurer, receiving the largest number of votes of any statewide candidate. Prior to government, she founded Point Judith Capital, a venture capital firm in Rhode Island, where she led the firm's health-care practice. Raimondo is an alumni fellow on the Yale Board of Trustees, a member of the Council of Foreign Relations, and serves on the board of the Truth Initiative. She earned her BA in economics from Harvard University, a PhD from Oxford University, and a JD from Yale Law School.

Laura J. Richardson, General, USA, Ret. served more than thirty-eight years in the U.S. Army, commanding units from the company to combatant command level and serving as the army's first female combat arms general officer. Most recently, she served as the commander of U.S. Southern Command, which is one of six Department of Defense geographic warfighting combatant commands, working directly for the president of the United States and the secretary of defense. Earlier in her career, she served as the commanding general of U.S. Army North, deputy commander of U.S. Forces Command, and deputy commanding general of the 1st Cavalry Division, and she ran the army's congressional liaison. In addition, she served as the military aide to the vice president. While commanding an assault helicopter battalion in the 101st Airborne Division, she deployed her unit to Iraq during the first year of Operation Iraqi Freedom, where she flew combat missions, and she served in Afghanistan in the International Security Assistance Force during Operation Enduring Freedom.

Peter L. Scher is the vice chairman of JPMorganChase, where he oversees several business and client functions for the firm, including the JPMorgan Chase Center for Geopolitics and Morgan Health. Scher led the firm's $200 million economic development investment in the revitalization of Detroit, which was featured by *60 Minutes*, *Fortune*, and Harvard Business School. Prior to JPMorganChase, Scher was the managing partner of the Washington, DC, office of law firm Mayer Brown. He spent nearly a decade in public service, including as U.S. special trade ambassador, chief of staff for the U.S. trade representative, and the U.S. secretary of commerce and staff director for the U.S.

Senate Committee on Environment and Public Works. Scher is a cofounder and former chairman of the Greater Washington Partnership and serves on the board of directors of CLEAR, health-care provider Centivo, the board of advisors for the Center for New American Security, the advisory board for the Stanford Emerging Technology Review, and a member of the Council on Foreign Relations. Scher received his BA from American University and his JD from American University's Washington College of Law.

Jonathan S. Spaner is a partner at McKinsey & Company's Washington, DC, office, where he focuses on issues involving integrated strategy, leadership, transitions, and crisis management in the public sector. He also serves as McKinsey's dean of faculty excellence. His recent client work includes helping a Fortune 300–scale public-sector organization develop a five-year portfolio strategy, leading the development of a technical roadmap for a public-sector organization to integrate biometric technologies into frontline missions, and developing the recruiting strategy and implementation plan for a Fortune 100–scale public-sector organization to meet financial management talent objectives. Previously, Spaner served as a commissioned officer in the U.S. Coast Guard for twenty-two years, holding the rank of captain and leading two major commands. On behalf of the State Department, Spaner led the U.S. delegation to the Arctic Council task force responsible for establishing the Arctic Economic Council. He also served as strategic policy advisor to a four-star general, director of port and cargo security on the White House staff, and as a White House fellow. Spaner is a former international affairs fellow at the Council on Foreign Relations and a former fellow at the German Marshall Fund of the United States. He holds a BS from the U.S. Merchant Marine Academy, where he served as the regimental commander, and an MBA from the Massachusetts Institute of Technology.

Jim Taiclet is chairman, president, and chief executive officer of Lockheed Martin. He became chairman in March 2021 after joining the company as president and CEO in June 2020. Taiclet has been a director on the Lockheed Martin board since January 2018. Prior to joining Lockheed Martin, he was chairman, president, and CEO of American Tower, one of the largest global real estate investment trusts. Taiclet guided the company's transformation from a primarily U.S. business to the only truly global player in its industry, with significant assets and operations in nineteen countries. Before he led American Tower, Taiclet served as president of AlliedSignal (subsequently Honeywell Aerospace Services). He also served as vice president, engine services, at Pratt & Whitney, where he was responsible for leading both military and commercial jet engine overhaul and repair. He is a member of the Business Council and an associate fellow of the American Institute of Aeronautics and

Astronautics and also serves on Mass General Brigham's board of directors. A distinguished graduate of the U.S. Air Force Academy, Taiclet earned bachelor's degrees in engineering and international relations. He also holds a master's degree from Princeton University, where he was awarded a fellowship at the Princeton School of Public and International Affairs.

Laura D'Andrea Tyson is a distinguished professor of the Haas School of Business's Graduate School at the University of California, Berkeley, and she chairs Berkeley's Blum Center for Developing Economies board of trustees. She previously served as cochair of California Governor Gavin Newsom's Council of Economic Advisors and is now a current member of the council. From July 2018 to December 2018, she served as interim dean of Berkeley Haas. Previously, she was the dean of London Business School from 2002 to 2006 and the dean of Berkeley Haas from 1998 to 2001. Tyson was a member of the U.S. Department of State Foreign Affairs Policy Board and a member of President Obama's Council on Jobs and Competitiveness and the President's Economic Recovery Advisory Board. She served as director of the National Economic Council from 1995 to 1996 and in the Clinton administration as the chair of the Council of Economic Advisers from 1993 to 1995. Tyson is the author of numerous books and reports on automation, artificial intelligence, the future of work, trade and industrial policy, sustainable investment, women's rights and economic performance, and the challenges of open strategic autonomy in Europe.

Dr. Sadek Wahba is chairman and managing partner of I Squared Capital, an independent global infrastructure investment manager based in Miami, Florida, with over $50 billion in assets under management. He has worked at Morgan Stanley as the CEO of Morgan Stanley Infrastructure and as an economist at the World Bank. He was a presidential appointment to the National Infrastructure Advisory Council, which advises the White House on reducing the physical and cyber risks to, and improving the security and resilience of, critical infrastructure in the United States. He was part of the expert committee on the World Economic Forum's first report on global infrastructure investments and was named Global Infrastructure Personality of the Year twice, as well as Global Infrastructure Personality of the Decade, by Private Equity International. He is a published author of economic research, and one of his publications was selected by MIT as one of its fifty most influential papers in the past fifty years. His book, *Build: Investing in America's Infrastructure*, published in 2024 and is in its second printing. Wahba is a senior fellow at the Development Research Institute of New York University, a foundation fellow of St. Antony's College, University of Oxford, and an advisor trustee of the American University in Cairo. He holds a BA in economics from the American University in Cairo, an MSc in economics from the London School of Economics and Political Science, and a PhD in economics from Harvard University.

Juan C. Zarate is the chairman and cofounder of the Center on Economic and Financial Power at the Foundation for Defense of Democracies (FDD), the cofounder and chair of Consilient, and the global comanaging partner at K2 Integrity. He is the lead independent trustee for Northwestern Mutual and a board member of the National Endowment for Democracy and Guardian Space Technology Solutions. He is a senior adviser at the Center for Strategic and International Studies; an advisor to the National Security Institute, the *Harvard National Security Journal*, and FDD's Center on Cyber and Technology Innovation; and a life member of the Council on Foreign Relations. Since 2014, Zarate has been an independent adviser to Coinbase, the largest U.S. virtual asset service provider. Zarate served on the boards of Boston Dynamics and Cambridge Quantum Computing North America, and was appointed twice by Pope Francis to the Vatican's Financial Information Authority Board. He was a visiting lecturer on law at Harvard Law School for eight years and is the author of books including *Treasury's War* and *Forging Democracy*. From 2005 to 2009, Zarate was the deputy assistant to President George W. Bush and deputy national security adviser for combating terrorism, the country's fifth counterterrorism czar. He was the first-ever assistant secretary of the treasury for terrorist financing and financial crimes, and he established the Treasury Department's Office of Terrorism and Financial Intelligence.

TASK FORCE OBSERVERS

Observers participate in Task Force discussions but are not asked to join the consensus. They participate in their individual, not institutional, capacities.

Heidi Crebo-Rediker is a senior fellow in the Center for Geoeconomic Studies at the Council on Foreign Relations, specializing in international political economy, U.S. economic competitiveness, economic security, and international finance. Crebo-Rediker served in the Obama administration as the State Department's first chief economist. She provided strategic advice to two secretaries of state on the integration of economics and finance with geopolitics to help launch "economic statecraft" in the administration. Her remit encompassed a range of foreign policy issues, both crisis-related and longer-term challenges and opportunities with economic drivers. Previously, Crebo-Rediker was chief of international finance and economics for the Senate Committee on Foreign Relations, following nearly two decades in Europe as a senior investment banker. In the Senate, she advised on international and domestic economic and financial issues, related, in particular, to the global financial crisis, the eurozone crisis and sovereign debt, the International Monetary Fund, multilateral development banks, and infrastructure investment. Her career started in energy merchant banking after working for one of the first U.S.-Russian joint ventures. She is a member of the Council on Foreign Relations and the Trilateral Commission and a previous member of the World Economic Forum's Global Agenda Council on the United States. She holds a BA from Dartmouth College and an MSc from the London School of Economics and Political Science.

Alexander M. Farman-Farmaian is a partner, vice chairman, and portfolio manager for Edgewood Management LLC. He joined in January 2006 as a partner and a member of the investment committee. He produces Economents™, an economic and financial markets update. Prior to Edgewood, Farman-Farmaian was a senior member of the portfolio management team at W.P. Stewart & Co. for nineteen years and chaired the investment oversight committee. He is a member of the Council on Foreign Relations, World Presidents' Organization, and Chief Executives Organization. Farman-Farmaian serves on the Museum of the City of New York's board of trustees, the Wikimedia Endowment board, and the executive advisory council of the George W. Bush Presidential Center. Additionally, Farman-Farmaian is a member of the Economics Club of New York and is on the board of Kalinat S.A. He received a BA in economics from Princeton University in 1987.

Matthew P. Goodman is distinguished fellow for geoeconomic studies at the Council on Foreign Relations. In 2023, he launched RealEcon: Reimagining American Economic Leadership, a CFR initiative that explores the U.S. role in the international economy. Prior to joining CFR in September 2023, Goodman was senior vice president for economics and Simon chair in political economy at the Center for Strategic and International Studies. From 2010 to 2012, he served as director for international economics on the National Security Council staff, helping the U.S. president prepare for global and regional summits, including for the Group of Twenty, Asia-Pacific Economic Cooperation, and East Asia Summit. Prior to serving in the White House, he was senior advisor to the undersecretary for economic affairs at the U.S. Department of State. Before joining the Obama administration in 2009, Goodman worked for five years at Albright Stonebridge Group, where he was managing director for Asia. From 2002 to 2004, he served at the White House as director for Asian economic affairs on the National Security Council staff. Prior to that, he spent five years at Goldman Sachs, heading the bank's government affairs operations in Tokyo and London. From 1988 to 1997, he worked as an international economist at the U.S. Treasury Department, including five years as financial attaché at the U.S. embassy in Tokyo. Goodman holds a BSc in economics from the London School of Economics and Political Science and an MA in international relations from the Johns Hopkins University School of Advanced International Studies.

Ziad Haider coleads McKinsey & Company's Geopolitics Practice. He counsels McKinsey's clients on how to build out capabilities and frameworks to understand, monitor, and mitigate geopolitical risks and is the firm's principal thought leader around building geopolitical resilience. Prior to joining McKinsey, Haider served in a range of senior roles relating to economic diplomacy and national security in the U.S. government, including as special representative for commercial and business affairs, engaging with and supporting U.S. businesses globally on market access and political and regulatory risks; as a member of the policy planning staff in the office of the secretary of state, conducting strategic planning around Asia and geoeconomics; as a White House fellow at the U.S. Department of Justice; and as a foreign policy legislative aide in the U.S. Senate. He previously practiced international law, specializing in sanctions and investment disputes, and worked with human rights organizations across Asia. His writing on international affairs has appeared in *Foreign Affairs*, *Foreign Policy*, and Nikkei Asia, among other leading publications. Ziad serves on the board of advisors of the International Rescue Committee, a leading global humanitarian and development organization, and is a nonresident senior advisor with the Center for Strategic and International Studies.

Shannon K. O'Neil is senior vice president, director of studies, and Maurice R. Greenberg chair at the Council on Foreign Relations, where she oversees the work of the more than six dozen fellows in the David Rockefeller Studies Program, as well as CFR's fourteen fellowship programs. She is a leading authority on global trade, supply chains, Mexico, and Latin America. Dr. O'Neil is the author of *The Globalization Myth: Why Regions Matter*, which chronicles the rise of three main global manufacturing and supply chain hubs and what they mean for U.S. economic competitiveness. She also wrote *Two Nations Indivisible: Mexico, the United States, and the Road Ahead*, which analyzes the political, economic, and social transformations Mexico has undergone over the past three decades and why they matter for the United States. She is a columnist for Bloomberg Opinion and a frequent guest on national broadcast news and radio programs. Dr. O'Neil has often testified before Congress and regularly speaks at global academic, business, and policy conferences. Dr. O'Neil has lived and worked in Mexico and Argentina. She was a Fulbright Scholar and a justice, welfare, and economics fellow at Harvard University, and has taught Latin American politics at Columbia University. Before turning to policy, Dr. O'Neil worked in the private sector as an equity analyst at Indosuez Capital and Credit Lyonnais Securities, and is the chair of the board of directors of the Tinker Foundation. She holds a BA from Yale University, an MA in international relations from Yale University, and a PhD in government from Harvard University.

Rebecca Patterson is a senior fellow at the Council on Foreign Relations. Patterson studies how politics and policy intersect with economic trends to drive financial markets. She is an independent director at Vanguard, a global asset manager with over $10 trillion in assets under management. Previously, Patterson was chief investment strategist for Bridgewater Associates. From 2012 through 2019, Patterson was chief investment officer of Bessemer Trust, managing $85 billion in client assets. Before Bessemer, Patterson spent over fifteen years at JPMorgan, where she worked as a researcher in the firm's investment bank in Europe, Singapore, and the United States; served as chief investment strategist in the asset management arm; and ran the private bank's global currency and commodity trading desk. Patterson transitioned to finance after several years as a journalist. Patterson has supported the Federal Reserve throughout her career, serving on the New York Federal Reserve's investor advisory and foreign exchange committees. She is a member of the Trilateral Commission and the Economic Club of New York and serves on the University of Florida's Investment Corporation Advisory Board. She chairs the Council for Economic Education and is a Bretton Woods Committee board member. She cohosted the first-ever television program on foreign-exchange markets: CNBC's *Money in Motion*. Patterson holds a BS in journalism from the University of Florida, an MA in international relations from the Johns Hopkins University School of Advanced International Studies, and an MBA from New York University.

Scott Sarlin was the national intelligence fellow at the Council on Foreign Relations from 2024 to 2025. He recently served as the deputy assistant director of national intelligence for mission performance, analysis, and collection at the Office of the Director of National Intelligence (ODNI). His office measures the performance of the intelligence community; facilitates the determination of intelligence priorities; represents mission equities in the intelligence community's budget, acquisition, and requirements processes; and leads functional management, collections, and analytic communities' interactions with the ODNI.

Anya Schmemann is managing director of the Task Force Program at the Council on Foreign Relations. At CFR, Schmemann has overseen numerous high-level Task Forces on a wide range of important national security topics, including cooperation in outer space, cybersecurity, China's Belt and Road initiative, pandemic preparedness, innovation, the future of work, Arctic strategy, nuclear weapons, climate change, immigration, trade policy, and internet governance, as well as on U.S. policy toward Afghanistan, Brazil, North Korea, Pakistan, Taiwan, and Turkey. She also served as managing director of communications at CFR for many years. She previously served as assistant dean for communications and outreach at American University's School of International Service and managed communications at Harvard Kennedy School's Belfer Center for Science and International Affairs, where she also administered the Caspian studies program. She coordinated a research project on Russian security issues at the EastWest Institute in New York and was assistant director of CFR's Center for Preventive Action in New York, focusing on the Balkans and Central Asia. She was a Truman national security fellow and is chair of the Global Kids DC advisory council. Schmemann received a BA in government and an MA in Russian studies from Harvard University.

Laura Taylor-Kale is a senior fellow for geoeconomics and defense at the Council on Foreign Relations. From 2023 to 2025, Taylor-Kale served as the first presidentially appointed, Senate-confirmed assistant secretary of defense for industrial base policy. In this role, she led all defense industrial strategy, investments, and planning, including the Defense Production Act and industrial base investments, supply chain resilience, small business programs, international defense industrial cooperation, economic security, and review of domestic mergers and acquisitions and foreign investments. She also developed the first-ever National Defense Industrial Strategy and Implementation Plan. Previously, Taylor-Kale was a fellow for innovation and economic competitiveness at CFR and an international affairs fellow, serving as deputy director and coauthor of the CFR Task Force report *The Work Ahead: Machines, Skills, and U.S. Leadership in the Twenty-First Century*. Earlier, as deputy assistant secretary for manufacturing in the Commerce Department's international trade administration, she led a trade policy team in breaking barriers in the export of U.S. manufactured goods globally. Taylor-Kale began her career in the U.S. Department of State as a career foreign service officer in South Asia and Africa and has worked at the World Bank and Overseas Private Investment Corporation. She is the founder and CEO of Strategic Capital Advisory LLC, an advisory firm working with venture capital, defense, and deep technology companies. She holds a BA from Smith College, an MPA from Princeton University's School of Public and International Affairs, an MBA from New York University's Stern School of Business, and a PhD in management science and engineering from Stanford University's School of Engineering.

Contributing CFR Staff

Dalia Albarrán
Associate Director,
Graphic Design

María Teresa Alzuru
Deputy Director,
Product Management

Sabine Baumgartner
Senior Photo Editor

Lucky Benson
Creative Director

Michael Bricknell
Data Systems Designer

Helena Kopans-Johnson
Research Associate, Trade Policy

Patricia Lee Dorff
Managing Director, Publications

Will Merrow
Associate Director,
Data Visualization

Caitlin Moran
Senior Editor, Publications

Anya Schmemann
Managing Director,
Task Force Program

Chelie Setzer
Deputy Director,
Task Force Program

Ishaan Thakker
Research Associate,
Geoeconomics

Katerina Viyella
Program Associate,
Task Force Program

Mali M. X
Director, Design
and User Experience

Contributing Interns

Liam Hurtt
Geoeconomics

Keyon Majidi
Task Force Program

Naya Patel
Task Force Program